BASIC MATHEMATICS

by Joseph P. Lerro, Jr.
DESIGN NEWS

Cahners Books, Inc.
221 Columbus Avenue,
Boston, Massachusetts 02116

Biography

About the author. Joseph P. Lerro, Jr. is a Dean's List graduate from Boston University where he received his B.S. degree in Mechanical/Aeronautical Engineering. He also has taken graduate courses in engineering at Northeastern University and Massachusetts Institute of Technology. While working as a project engineer with Avco Corp., where he was responsible for all facets of engineering from design and testing through fabrication and final installation of the completed design, Joe was awarded the 1968 "Engineer of the Year" award. The author is currently a full-time member of the *Design News* staff, a part-time teacher at Franklin Institute of Boston where he instructs students in such engineering subjects as structural analysis and advanced fluid dynamics and is president of his own company where he does part-time engineering consulting work. He has two patents pending approval and is the author of another book entitled "Metric System." The 1974 Directory of Metric Consultants, recently published by *Metric News,* lists Joe as a Metric Consultant.

Preface

This book on basic mathematics is intended for engineers, physicists, chemists, draftsmen, technicians, teachers, high school and college students and all others who need a good handbook of the basic laws of mathematics.

The purpose of this book is multifold. It can be used by people with no knowledge of mathematics who are interested in learning the basic laws and people who are interested in a general review. And probably the most important reason for the book, is the person who has to make use of mathematics in their everyday functions. This book eliminates the countless hours of searching through dozens of reference books to find needed equations and then enduring the drudgery of cranking out the necessary calculations. Basic Mathematics provides a single collection of the most-used laws, equations and formulas, all boiled down and expressed in simple terms, graphs, charts, diagrams, figures, tables and nomograms to give quick, accurate results.

The following areas of mathematics are covered: algebra, trigonometry, plane geometry, solid geometry, analytical geometry, integral calculus, differential calculus and mensuration. These areas are presented in the book by dividing it into the following six major sections:

1. **General Information**-This section is intended to give the reader a basic understanding of the areas of mathematics that are covered in the book. It presents to the reader a solid foundation necessary to proceed through the book.

2. **Area Information**-The majority of area information is provided in the form of equations and nomogramical solution of these equations. The equations can be used by themselves or the nomogram may be used as a short cut.

3. **Volume Information**-Volume information is also provided in the form of equations and nomograms. The format and usage of this section is similar to the Area Information section.

4. **Angle Information**-This section consists mostly of charts and tables to solve for the different angular relationships. Detailed analysis is given of the triangle and the relationship between its sides and angles.

5. **Equation Information**-Many varied types of equations are solved for in this section ranging from quadratic and cubic to complex and "unsolvable". Differential and integral calculus equations with their related axioms are also presented.

6. **Linear Information**-Articles ranging from finding the square and cube root of a number, use of the pythagorean theorem and solution of geometric problems are presented in this section in the form of charts, diagrams and nomograms.

The study of mathematics is a very interesting subject and I hope this book will add to the users' interest and understanding of the subject. The intent of the book is to present basic, practical mathematical calculations and not include any of the higher advanced mathematical methods. If these goals have been accomplished, the book will have found itself a place in the mathematical world.

Joseph P. Levro Jr.

Contents

Contents

Basic laws of solid geometry

Introduction.
The following laws, definitions and equations include those most often used in basic mathematical calculations. For convenient reference, a topical index is given below. The numbers refer to the items in this article.

1. POINT

A point is that undefined geometric element having position but no magnitude, dimensions, length, breadth or thickness.

●

2. LINE

A line is length without breadth or thickness.

———————

3. SURFACE

A surface is an extension or a figure of two dimensions—length and breadth, but without thickness.

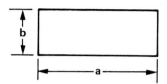

4. BODY

A body or solid is a figure of three dimensions—length, breadth and depth or thickness.

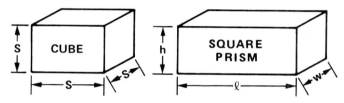

5. LINES

Lines are either straight or curved, or a combination of these two.

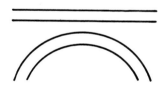

6. STRAIGHT LINE

A straight line lies all in the same direction between its extremities and is the shortest distance between two points. (Note: when a line is simply mentioned, it means a straight line).

7. CURVE

A curve continually changes its direction between its extreme points.

8. FOUR TYPES OF LINES

Lines are: Parallel
Oblique
Perpendicular
Tangential

9. PARALLEL LINES

Parallel lines are always at the same perpendicular distance and they never meet though ever so far apart produced.

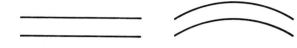

10. OBLIQUE LINES

Oblique lines change their distance and would meet if produced on the side of the least distance.

11. PERPENDICULAR LINES

One line is perpendicular to another when it inclines not more on the one side than the other or when the angles on both sides of it are equal.

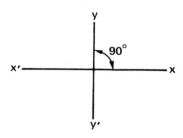

12. TANGENT, TANGENTIAL LINE OR CIRCLE

A line or circle is tangential or is a tangent to a circle or the other curve, when it touches it without cutting, although both are produced. Also, tangent is the meeting of a curve or surface at two or more consecutive points and hence having there the same direction as the curve or surface.

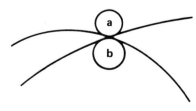

13. ANGLE

An angle is the inclination or opening of two lines having different directions and meeting in a point.

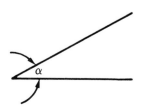

14. ANGLES
Angles are: Right
     Oblique
     Acute
     Obtuse

15. RIGHT ANGLE
A right angle is that which is made by one line perpendicular to another or when the angles on each side are equal to one another they are right angles.

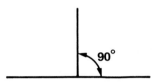

16. OBLIQUE ANGLE
An oblique angle is that which is made by two oblique lines and is either less or greater than a right angle.

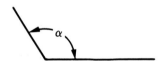

17. ACUTE ANGLE
An acute angle is less than a right angle.

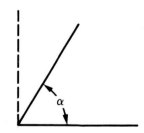

18. OBTUSE ANGLE
An obtuse angle is greater than a right angle.

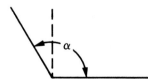

19. SURFACES
Surfaces are either plane or curved.

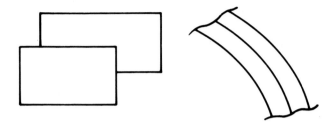

20. PLANE SURFACE OR A PLANE
A plane surface or a plane is that which a straight line may every way coincide. If the line touches the plane in two points, it will touch it in every point; if not, it is curved.

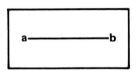

21. PLANE FIGURES
Plane figures are bounded either by straight lines or curves.

22. PLANE FIGURES BOUNDED BY STRAIGHT LINES
Plane figures that are bounded by straight lines are specifically named according to their number of sides or of their angles. They have as many sides as angles, the least being three.

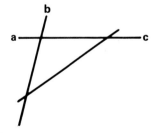

23. TRIANGLE
A triangle has three sides and three angles and it receives particular denominations from the relations of its sides and angles.

24. EQUILATERAL TRIANGLE
An equilateral triangle is that whose three sides are all equal.

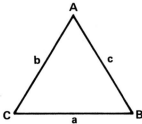

25. ISOSCELES TRIANGLE

An isosceles triangle is that which has two sides equal.

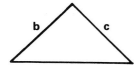

26. SCALENE TRIANGLE

A scalene triangle is that whose three sides are all unequal.

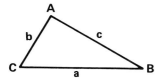

27. RIGHT-ANGLE TRIANGLE

A right-angle triangle is that which has one right angle.

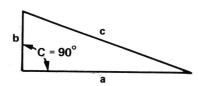

28. OBLIQUE-ANGLE TRIANGLE

Other triangles are oblique-angled and are either obtuse or acute.

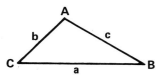

29. OBTUSE TRIANGLE

An obtuse triangle has one obtuse angle.

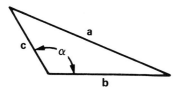

30. ACUTE-ANGLE TRIANGLE

An acute-angle triangle has all of its three angles acute.

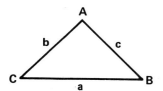

31. QUADRANGLE OR A QUADRILATERAL

A figure of four sides and four angles is called a quadrangle or a quadrilateral.

32. PARALLELOGRAM

A parallelogram is a quadrilateral that has both its pairs of opposite sides parallel and is known by the following names: rectangle, square, rhomboid and rhombus.

33. RECTANGLE

A rectangle is a parallelogram having a right angle.

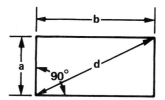

34. SQUARE

A square is an equilateral rectangle having its length and breadth equal, all of its sides equal and all its angles equal.

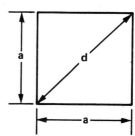

35. RHOMBOID

A rhomboid is an oblique-angle parallelogram.

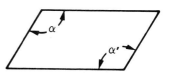

36. RHOMBUS

A rhombus is an equilateral rhomboid having all its sides equal but its angles oblique.

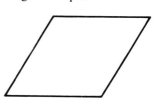

37. TRAPEZIUM
A trapezium is a quadilateral whose opposite sides are not parallel.

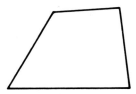

38. TRAPEZOID
A trapezoid has only one pair of opposite sides parallel.

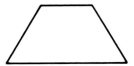

39. DIAGONAL
A diagonal is a line joining any opposite angles of a quadrilateral.

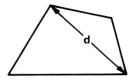

40. POLYGONS
Plane figures that have more than four sides are generally called polygons and they receive other names according to the number of their sides or angles.

41. PENTAGON, HEXAGON, HEPTAGON, OCTAGON, NONAGON, DECAGON, UNDEGON AND DODEGON
A pentagon is a polygon of five sides; hexagon has six; heptagon has seven; octagon has eight; nonagon has nine; decagon has ten; undegon has eleven and dodegon has twelve.

42. REGULAR POLYGON
A regular polygon has all its sides and all its angles equal. If they are not both equal, the polygon is irregular.

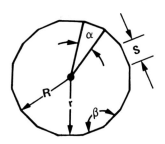

43. REGULAR FIGURE
An equilateral triangle is also a regular figure of three sides, the square is one of four sides. The former being also called a trigon and the latter a tetragon.

44. EQUILATERAL FIGURE
Any figure is equilateral when all its sides are equal and it is equiangular when all its angles are equal. When both of these are equal, it is a regular figure.

45. CIRCLE
A circle is a plane figure bounded by a curved line called a circumference, which is equidistant from a point within called a center. The circumference itself is often called a circle and also the periphery.

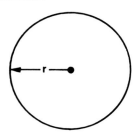

46. RADIUS OF A CIRCLE
The radius of a circle is a line drawn from the center to the circumference.

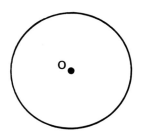

47. DIAMETER OF A CIRCLE
The diameter of a circle is a line drawn through the center and terminating at the circumference on both sides.

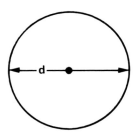

48. ARC OF A CIRCLE
An arc of a circle is any part of the circumference.

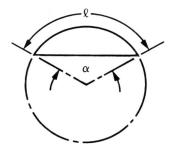

49. CHORD
A chord is a straight line joining the extremities of an arc.

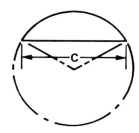

50. SEGMENT
A segment is any part of a circle bounded by an arc and its chord.

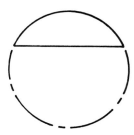

51. SEMICIRCLE
A semicircle is half the circle or a segment cut off by a diameter. The half circumference is sometimes called the semicircle.

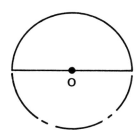

52. SECTOR
A sector is any part of a circle that is bounded by an arc and two radii drawn to its extremities.

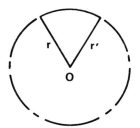

53. QUADRANT OR QUARTER OF A CIRCLE
A quadrant or quarter of a circle is a sector having a quarter of the circumference for the arc and its two radii are perpendicular to each other. A quarter of a circumference is sometimes called a quadrant.

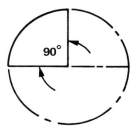

54. HEIGHT
The height or altitude of a figure is a perpendicular drawn from an angle or its vertex, to the other side called the base.

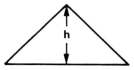

55. HYPOTENUSE
In a right-angle triangle, the side opposite the right angle is called the hypotenuse and the other two sides are called the legs. They are sometimes called the base and perpendicular or side opposite and side adjacent.

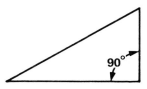

56. ANGULAR POINT
When an angle is denoted by three letters, of which one stands at the angular point and the other two on the two sides, that which stands at the angular point is read in the middle.

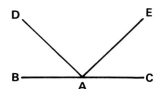

57. CIRCUMFERENCE

The circumference of every circle is divided into 360 equal parts called degrees; each degree into 60 minutes; each minute into 60 seconds. Thus, a semicircle contains 180 degrees and a quadrant 90 degrees.

58. MEASURE OF AN ANGLE

The measure of an angle is an arc of any circle contained between the two lines that form that angle—the angular point being the center. It is measured by the number of degrees contained in an arc.

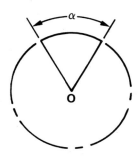

59. LINES OR CHORDS

Lines or chords are said to be equidistant from the center of the circle when perpendiculars drawn to them from the center are equal.

60. GREATER PERPENDICULAR

Lines or chords are said to be equidistant from the center of the circle when perpendiculars drawn to them from the center are equal and the straight line on which the greater perpendicular falls is said to be furthest from the center. The length ab of the straight line is greater than the other two lines.

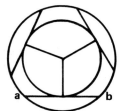

61. ANGLE IN A SEGMENT

An angle in a segment is that which is contained by two lines, drawn from any point in the opposite or supplementary part of the circumference to the extremities of the arc, and containing the arc between them.

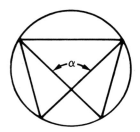

62. ANGLE ON A SEGMENT

An angle on a segment or an arc is that which is contained by two lines, drawn from any point in the opposite or supplementary part of the circumference to the extremeties of the arc, and containing the arc between them.

63. ANGLE AT THE CIRCUMFERENCE

An angle at the circumference is that whose angular point or summit is anywhere in the circumference. Also, an angle at the center is that whose angular point is at the center.

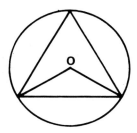

64. INSCRIBED CIRCLE

A straight-lined figure is inscribed in a circle or the circle circumscribes it, when all the angular points of the figure are in the circumference of the circle.

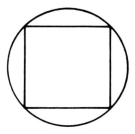

65. CIRCUMSCRIBED CIRCLE

A straight-lined figure circumscribes a circle or the circle is inscribed in it, when all the sides of the figure touch the circumference of the circle.

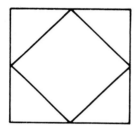

66. STRAIGHT-LINE FIGURE INSCRIBED IN ANOTHER
One straight-lined figure is inscribed in another or the latter circumscribes the former, when all the angular points of the former are placed in the sides of the latter.

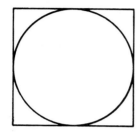

67. SECANT
A secant is a line that cuts a circle lying partly within and partly outside of it.

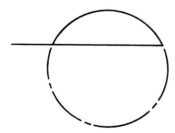

68. MUTUALLY EQUILATERAL
Two triangles or other straight-lined figures are said to be mutually equilateral, when all sides of one are equal to the corresponding sides of the other. They are said to be equiangular when the angles of one are respectively equal to those of the other.

69. IDENTICAL FIGURES
Identical figures are mutually equilateral and equiangular—all sides and angles of one are respectively equal to all sides and angles of the other. If one figure were applied to or laid upon the other, the two become one and the same figure.

70. SIMILAR FIGURES
Similar figures are those that have all angles of one equal to all angles of the other and the sides about the equal angles proportional.

71. PERIMETER
The perimeter of a figure is the sum of all its sides taken together.

72. ALTITUDE
The perpendicular distance from the base of a figure to the summit is the altitude.

73. PERPENDICULAR
A line at right angles to the plane of the horizon or to another line is called the perpendicular.

74. DEGREE
The degree is the 360th part of the circumference or of a round angle—its symbol is °.

75. MINUTE
A minute is the sixtieth part of a degree—its abbreviation is m or min.

76. SECOND
A second is the sixtieth part of a minute of a degree—its abbreviation is s or sec.

77. SOLID
A solid is a magnitude that has three dimensions—length, breadth and thickness. Also, a solid is a part of space bounded on all sides such as a cube or sphere.

78. BASE
The base is the line or surface constituting the part of a figure on which it is supposed to stand.

79. VECTOR
The vector is a complex entity representative of a directed magnitude such as a force or velocity, and represented by any of a system of equal and parallel line segments.

80. VERTEX
The vertex is found in any figure having a base and it is the point opposite to and farthest from the base or the top. Also, the vertex is the terminating point of some particular line or lines in a figure or curve, such as where the sides of an angle meet or where a curve or surface meets the axis.

81. SCALAR
In vector analysis, scalar is an undirected quantity and it is a quantity fully described by a number.

82. ORTHOCENTER
The altitudes meet in a point called the orthocenter that is usually designated by the letter o or O.

83. EXTERIOR ANGLE
An exterior angle equals the sum of the two opposite interior angles.

84. MEDIANS
The medians—joining each vertex with the middle point of the opposite side, meet in the center of gravity which trisects each median.

85. CENTER OF GRAVITY
The center of gravity is that point in a body about which all the parts of a body exactly balance each other.

86. BISECTORS OF THE ANGLES
The bisectors of the angles meet in the center of the inscribed circle.

87. RADICAL AXIS
The radical axis of two circles is a straight line, such that the tangents drawn from any point of this line to the two circles, are of equal length.

88. DIHEDRAL ANGLES
The dihedral angle between two planes is measured by a plane angle formed by two lines, one on each plane perpendicular to the edge.

89. SPHERE
The sphere is a body of space bounded by one surface, all points of which are equally distant from a point within called its center.

90. ANNULUS
The annulus is a ring and it is also a ringlike part, structure or space.

91. VOLUME
Volume is the space occupied as measured by cubic units.

92. PROPOSITION
A proposition is something that is either proposed to be done or to be demonstrated, and is either a problem or a theorem.

93. PROBLEM
A problem is something proposed to be done.

94. THEOREM
A theorem is something proposed to be demonstrated.

95. LEMMA
A lemma is something that is premised or demonstrated in order to render what follows more closely.

96. COROLLARY
A corollary is a constant truth gained immediately from some preceding truth or demonstration.

97. SCHOLIUM
A scholium is a remark or observation made upon something going before it.

98. POSTULATE
A postulate is something required to be done.

99. HYPOTHESIS
An hypothesis is a supposition assumed to be true, in order to argue from or base upon it the reasoning and demonstration of some proposition.

100. DEMONSTRATION
Demonstration is the collection of several arguments and proofs, laying them together in proper order to show the truth of the proposition under consideration.

101. DIRECT, POSITIVE OR AFFIRMATIVE DEMONSTRATION
A direct, positive or affirmative demonstration is that which concludes with the direct and certain proof of the proposition in hand. It is sometimes called an ostensive demonstration.

102. INDIRECT OR NEGATIVE DEMONSTRATION
That which shows a proposition to be true by proving some absurdity would necessarily follow if the proposition advanced were false. This is sometimes called Reductio ad Absurdum.

103. METHOD
Method is the art of disposing a train of arguments in a proper order to investigate either the truth or falsity of a proposition or to demonstrate to others when it has been found out. This is either analytical or synthetical.

104. ANALYSIS
Analysis or the analytic method, is the art or mode of finding out the truth of a proposition, by first assuming the thing to be done and reasoning back step by step till we arrive at some truth. This is commonly called method of invention or resolution.

105. SYNTHESIS

Synthesis or the synthetic method, is the searching out of the truth by first laying down some simple and easy principle, and pursuing the consequences flowing from it till we arrive at the conclusion. This is also called the method of composition.

106. RESOLUTION

Resolution is the act or process of resolving or reducing to simpler form. Also, its result, answer or solution.

107. AXIOMS

A proposition regarded as a self-evident truth.

AXIOM 108

Things that are equal to the same thing are equal to each other.

AXIOM 109

When equals are added to equals the wholes are equal.

AXIOM 110

When equals are taken from equals the remainders are equal.

AXIOM 111

When equals are added to unequals the wholes are unequal.

AXIOM 112

When equals are taken from unequals the remainders are unequal.

AXIOM 113

Things that are double of the same thing or equal things are equal to each other.

AXIOM 114

Things that are halves of the same things are equal.

AXIOM 115

Every whole is equal to all of its parts taken together.

AXIOM 116

Things that coincide or fill the same space are identical or mutually equal in all their parts.

AXIOM 117

All right angles are equal to each other.

AXIOM 118

Angles that have equal measures or arcs are equal.

THEOREM 119

When two triangles have two sides and the included angle in one equal to two sides and the included angle in the other, the triangles will be identical or equal in all respects. If side AC equals side DF and side BC equals side EF and angle C equals angle F, the two triangles will be equal.

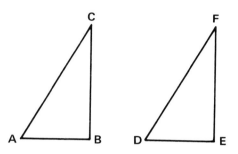

THEOREM 120

When two triangles have two angles and the included side in one equal to two angles and the included side in the other, the triangles are identical or have their other sides and angles equal. Let the two triangles ABC and DEF have the angle A equal to the angle D and angle B equal to angle E and side AB equal to side DE, then these two triangles will be identical.

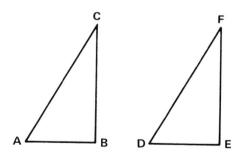

THEOREM 121

In an isosceles triangle the angles at the base are equal or if a triangle has two sides equal, their opposite angles will also be equal. If triangle ABC has side AC equal to side BC, then angle B will be equal to angle A.

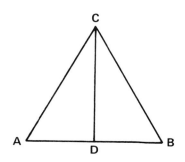

THEOREM 122

When a triangle has two of its angles equal, the sides opposite them are also equal. If triangle ABC has angle A equal to angle B, it will also have side AC equal to side BC.

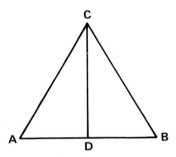

THEOREM 123

When two triangles have all three sides in one equal to all three sides in the other, the triangles are identical or also have their three angles equal—each to each. Let the two triangles ABC and ABD have their three sides respectively equal; side AB equal to AB, AC equal to AD and BC equal to BD. Then, the two triangles will be identical or have their angles equal, namely those angles that are opposite the equal sides:

Angle BAC = Angle BAD
Angle ABC = Angle ABD
Angle C = Angle D

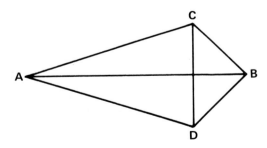

THEOREM 124

When one line meets another, the angles that it makes on the same side of the other line are both equal to two right angles. Let line AB meet line CD, then the two angles ABC and ABD taken together will be equal to two right angles.

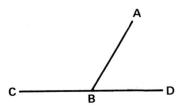

THEOREM 125

When two lines intersect each other, the opposite angles are equal. Let the two lines AB and CD intersect at the point E, then angle AEC will be equal to angle BED, and angle AED will be equal to angle CEB.

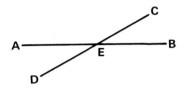

THEOREM 126

When one side of a triangle is produced, the outward angle is greater than either of the two inward opposite angles. Let ABC be a triangle having side AB produced to D, then the outward angle CBD will be greater than either of the inward opposite angles A or C.

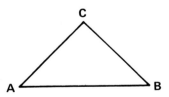

THEOREM 127

The greater side of every triangle is opposite the greater angle. Let ABC be a triangle having side AB greater than side AC, then angle ACB opposite the greater side AB, will be greater than angle ABC opposite the smaller side AC.

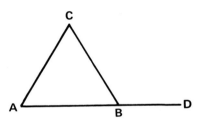

THEOREM 128

The sum of any two sides of a triangle is greater than the third side. Let ABC be a triangle, then the sum of any two of its sides will be greater than the third side. For example, AC + CB is greater than AB.

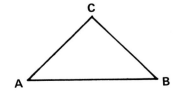

THEOREM 129

The difference of any two sides of a triangle is less than the third side. Let ABC be a triangle, then the difference of any two sides such as AB − AC will be less than the third side BC.

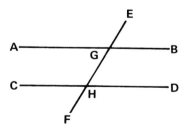

THEOREM 130

When a line intersects two parallel lines, it makes the alternate angles equal to each other. Let line EF cut two parallel lines AB and CD, then angle AGH will be equal to the alternate angle GHD.

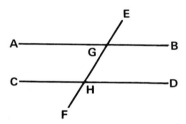

THEOREM 131

When a line cutting two other lines makes the alternate angles equal to each other, those two lines are parallel. Let line EF cutting the two lines AB and CD make alternate angles AGH and DHG equal to each other, then AB will be parallel to CD.

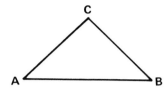

THEOREM 132

When a line cuts two parallel lines, the outward angle is equal to the inward opposite one on the same side and the two inward angles on the same side are both equal to two right angles. Let line EF cut the two parallel lines AB and CD, then the outer angle EGB will be equal to the inner opposite angle GHD on the same side of line EF, and the two inner angles BGH and GHD will be equal to two right angles.

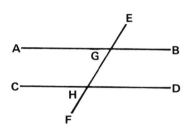

THEOREM 133

Lines that are parallel to the same line are parallel to each other. Let lines AB and CD be parallel to line EF, then lines AB and CD will be parallel to each other.

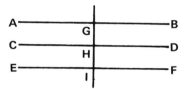

THEOREM 134

When one side of a triangle is produced, the outer angle is equal to both the inner opposite angles taken together. Let side AB of triangle ABC be produced to D, then the outer angle CBD will be equal to the sum of the two inner opposite angles A and C.

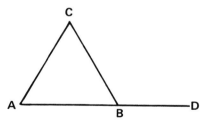

THEOREM 135

In any triangle, the sum of all three angles is equal to two right angles. Let ABC be any plane triangle, then the sum of the three angles A + B + C is equal to two right angles.

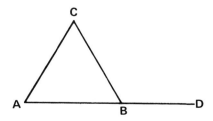

THEOREM 136

In any quadrangle, the sum of all the four inward angles is equal to four right angles. Let ABCD be a quadrangle, then the sum of the four inward angles A + B + C + D is equal to four right angles.

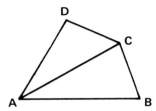

THEOREM 137

In any figure, the sum of all the inward angles taken together, is equal to twice as many right angles, wanting four, as the figure has sides. Let ABCDE be any figure, then the sum of all its inward angles A + B + C + D + E is equal to twice as many right angles, wanting four, as the figure has sides.

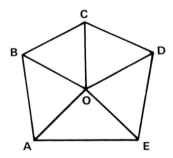

THEOREM 138

When every side of any figure is produced out, the sum of all the outer angles thereby made is equal to four right angles. Let A, B, C, D and E be the outer angles of any polygon made by producing all the sides, then the sum A + B + C + D + E of all those outer angles will be equal to four right angles.

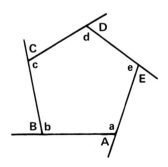

THEOREM 139

A perpendicular is the shortest line that can be drawn from a given point to an indefinite line. Of any other lines drawn from the same point, those that are nearest the perpendicular are less than those more remote.

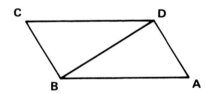

THEOREM 140

The opposite sides and angles of any parallelogram are equal to each other and the diagonal divides it into two equal triangles. Let ABCD be a parallelogram with a diagonal of BD, then its opposite sides and angles will be equal to each other and the diagonal BD will divide it into two equal parts or triangles.

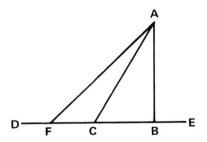

THEOREM 141

Every quadrilateral whose opposite sides are equal or has its opposite sides parallel, is a parallelogram. Let ABCD be a quadrangle having the opposite sides equal, side AB equal to DC and side AD equal to BC, then these equal sides will also be parallel and the figure is a parallelogram.

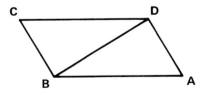

THEOREM 142

Rectangles that are contained by equal lines are equal to each other. Let BD and FH be two rectangles having sides AB and BC equal to sides EF and FG—each to each, then the rectangle BD will be equal to the rectangle FH.

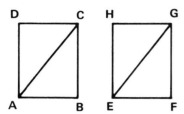

THEOREM 143
The sum of all the rectangles contained under one whole line and several parts of another line—any way divided, is equal to the rectangle contained under two whole lines. Let AD be one line and AB the other line divided into parts AE, EF and FB, then the rectangle contained by AD and AB will be equal to the sum of the rectangles of AD and AE, AD and EF, and AD and FB.

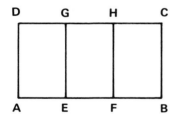

THEOREM 144
In any parallelogram the two diagonals bisect each other and the sum of their squares is equal to the sum of the squares of all four sides of the parallelogram. Let ABCD be a parallelogram whose diagonals intersect each other at E, then AE equals EC and BE equals ED, and the sum of the squares of AC and BD will be equal to the sum of the squares of AB, BC, CD and DA.

$$AE = EC$$
$$BE = ED$$
$$(AC)^2 + (BD)^2 = (AB)^2 + (BC)^2 + (CD)^2 + (DA)^2$$

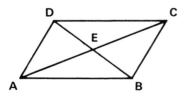

THEOREM 145
If a line drawn through or from the center of a circle bisects a chord, it will be perpendicular to it or if it is perpendicular to the chord it will bisect both the chord and arc of the chord. Let AB be any chord in a circle and CD a line drawn from the center C to the chord. Then, if the chord is bisected at point D, CD will be perpendicular to AB.

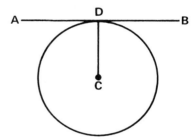

THEOREM 146
A line perpendicular to the extremity of a radius is a tangent to the circle. Let line AB be perpendicular to the radius CD of a circle, then AB will touch the circle at point D only.

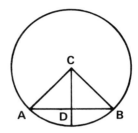

THEOREM 147
An angle in a semicircle is a right angle. If ADC is a semicircle then any angle D in that semicircle is a right angle.

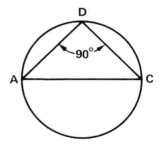

THEOREM 148
The angle formed by a tangent to a circle and a chord drawn from the point of contact is equal to the angle in the alternate segment. If AB is tangent to a circle and EC a chord and D any angle in the alternate segment EDC, then angle D will be equal to angle BEC made by the tangent and chord of the arc EC.

15

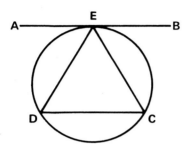

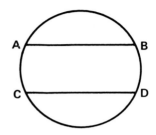

THEOREM 149
The sum of any two opposite angles of a quadrangle inscribed in a circle is equal to two right angles. Let ABCD be any quadrilateral inscribed in a circle, then the sum of the two opposite angles A and C or B and D, will be equal to two right angles.

THEOREM 152
When a tangent and chord are parallel to each other, they intercept equal arcs. Let tangent AC be parallel to chord DF, then arcs BD and BF will be equal.

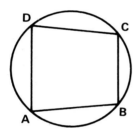

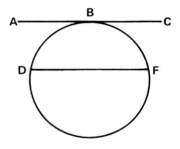

THEOREM 150
If any side of a quadrilateral inscribed in a circle be produced out, the outer angle will be equal to the inner opposite angle. If side AB of quadrilateral ABCD inscribed in a circle is produced to E, the outer angle DAE will be equal to the inner opposite angle C.

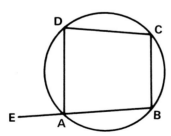

THEOREM 151
Any two parallel chords intercept equal arcs. Let two chords AB and CD be parallel, then arcs AC and BD will be equal.

Mathematical charts

Introduction.

The following mathematical charts represent common geometric figures and their related equations. The geometric figures are alphabetically arranged and the following example will illustrate the use of the charts.

Example 1:

Given a right-circular cone that has a radius of 2.0 inches and height of 5.0 inches, determine the shape of the right-circular cone and its volume.

Solution 1:

Going through the mathematical tables we locate a right-circular cone. Its shape is as follows:

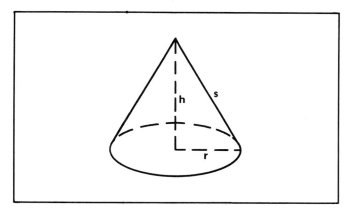

The volume of a right-circular cone is given by the following equation:

$$V = \frac{1}{3}\pi r^2 h$$

Substituting into the equation we arrive at:

$$V = \frac{1}{3}(3.1416)(2.0^2)(5.0)$$

$$V = \frac{1}{3}(3.1416)(4.0)(5.0)$$

$$V = \frac{62.8}{3}$$

$$V = 20.9 \text{ cu in}$$

Note:

Throughout the mathematical charts you will find notes called out in certain areas where I felt a further explanation was necessary. These notes appear at the end of the mathematical tables and supply the additional information necessary to completely understand the charts.

Name of Figure	ACUTE-ANGLE TRIANGLE	ANCHOR RING	ANNULUS	ANY PRISM OR CYLINDER
Classification	Plane Figure	Solid	Plane Figure	Solid
Kind of Figure				
Area (A)	$A = bh/2$ $A = \dfrac{b}{2}\sqrt{a^2 - \left(\dfrac{a^2 + b^2 - c^2}{2b}\right)^2}$ (See Note 1)	$A = 4\pi^2 cr = 39.478cr$ $A = \pi^2 Dd = 9.8696Dd$ (See Note 2)	$A = \pi(R^2 - r^2)$ $A = \pi(D^2 - d^2)/4$ $A = 2\pi R'b$ (Ref. Circular Ring) (See Note 11)	B = area of base N = area of normal section
Volume (V)		$V = 2\pi^2 cr^2 = 19.739cr^2$ $V = \dfrac{\pi^2}{4}Dd^2 = 2.4674Dd^2$		$V = Bh$ $V = N\ell$
Height (h)	$h = 2A/b$			$h = V/B$
Sides (a, b or c)	$b = 2A/h$	$c = \dfrac{A}{4\pi^2 r}$	$b = R - r$ $b = A/2\pi R'$	$B = h/V$
Radius (r, r', R, R')		$r = \dfrac{A}{4\pi^2 c}$	$R' = \dfrac{1}{2}(R + r)$	
Minor Dia. (d), Major Dia. (D)		$d = \sqrt{\dfrac{V}{\frac{\pi^2 D}{4}}} = \sqrt{\dfrac{V}{2.467D}}$ $D = \dfrac{V/\pi^2 d^2}{4} = V/2.4674d^2$		
Length of Element (ℓ)				$\ell = LA/Q = V/N$ where Q = perimeter of normal section
Lateral Area (LA)				$LA = Q\ell$

Name of Figure	ANY PYRAMID OR CONE	ANY PYRAMID OR CONICAL FRUSTUM	ANY QUADRILATERAL	ANY SPHERICAL SEGMENT (ZONE)
Classification	Solid	Solid	Plane Figure	Solid
Kind of Figure				
Area (A)			$A = \frac{1}{2}D_1D_2 \sin \mu$	
Volume (V)	$V = 1/3$ (area of base) X (altitude) $v = 1/3\,Bh$	$V = 1/3h(B + \sqrt{BB'} + B')$ $V = 1/3hb[1 + (P'/P) + P'P)^2]$ (See Note 3)		$V = \frac{1}{6}\pi h(3a^2 + 3a_1^2 + h^2)$ (See Note 4)
Height (h)	$h = V/\frac{1}{3}B$	$h = \dfrac{V}{1/3(B + \sqrt{BB'} + B')}$ $h = \dfrac{V}{\frac{1}{3}B[1 + (P'/P) + P'/P)^2]}$		$h = \dfrac{LA}{2\pi r}$
Sides (a, b or c)	$B = V/\frac{1}{3}h$		D_1 and D_2 are the diagonals and m = distance between midpoints of D_1 and D_2. Also, $a^2 + b^2 + c^2 + d^2 = D_1^2 + D_2^2 + 4m^2$	c = slant height of frustum $c = \sqrt{h^2 + (a - a_1)^2}$
Radius (r, r', R, R')				$r = \dfrac{LA}{2\pi h}$
Minor Dia. (d), Major Dia. (D)				
Length of Element (ℓ)			length of D_2 $= A/\frac{1}{2}D_1 \sin \mu$ length of D_1 $= A/\frac{1}{2}D_2 \sin \mu$	
Lateral Area (LA)	LA of regular pyramid = $\frac{1}{2}$ (perimeter of base) X (slant height)			$LA = 2\pi rh$

19

Name of Figure	ANY TRIANGLE	ANY TRUNCATED PRISM OR CYLINDER	ANY UNGULAR OF A RIGHT CIRCULAR CYLINDER	BARREL
Classification	PLANE FIGURE	SOLID	SOLID	SOLID
Kind of Figure				
Area (A)	$A = \frac{1}{2}$ base × Altitude $A = \frac{1}{2} ah = \frac{1}{2} ab \sin C$ $A = rs = abc/4R$ $A = \sqrt{s(s-a)(s-b)(s-c)}$ (See Note 5)	(See Note 6)		
Volume (V)		$V = N\ell$	$V = H(\frac{2}{3}a^3 \pm cB)/(r \pm c)$ $V = H[a(r^2 - \frac{1}{3}a^2) \pm$ $r^2c \text{ rad } \gamma/(r \pm c)$	$V = \frac{1}{12}\pi h(2D^2 + d^2)$ $V = 0.262\, h(2D^2 + d^2)$ (See Note 7) $V = 0.209\, h(2D^2 + Dd + \frac{3}{4}d^2)$ (See Note 8)
Height (h)			$H = \dfrac{V}{(2/3a^3 \pm cB)/(r \pm c)}$ $H = \dfrac{V}{[a(r^2 - \frac{1}{3}a^2) \pm r^2c \text{ rad } \gamma]}$ $/(r \pm c)$	$h = \dfrac{V}{1/12\pi(2D^2 + d^2)}$ $h = \dfrac{V}{0.262(2D^2 + d^2)}$ (see Notes 7 & 8)
Sides (a, b or c)	$s = \frac{1}{2}(a + b + c)$ $t = \frac{1}{2}(m_1 + m_2 + m_3)$ s and t are merely summations and cannot be shown graphically.			
Radius (r, r', R, R')	$r = \sqrt{(s-a)(s-b)(s-c)}/s$ = radius of inscribed circle $R = \frac{1}{2}a/\sin A = \frac{1}{2}b/\sin B$ $= \frac{1}{2}c/\sin C$ = radius of circumscribed circle			
Minor Dia. (d), Major Dia. (D)				
Length of Element (ℓ)		$\ell = V/N$ $k = LA/Q$ (See Note 6)		
Lateral Area (LA)		$LA = Qk$ $N = V/\ell$ $Q = LA/k$ (See Note 6)	$LA = H(2ra \pm cs)/(r \pm c)$ $LA = 2rH(a \pm c \text{ rad } \gamma)/$ $(r \pm c)$	

20

Name of Figure	CIRCLE	CIRCULAR RING	CIRCULAR RING SECTOR	CIRCULAR SECTOR
Classification	PLANE FIGURE	PLANE FIGURE	PLANE FIGURE	PLANE FIGURE
Kind of Figure				
Area (A)	$A = \pi r^2 = 3.1416 r^2$ $A = 0.7854\, d^2$ Circumference $C = 2\pi r = 6.2832 r$ $C = 3.1416 d$	$A = 3.1416(R^2 - r^2)$ $A = 0.7854(D^2 - d^2)$ (Ref. Annulus) (See Note 11)	$A = 0.00873\, \alpha(R^2 - r^2)$ $A = 0.00218\, \alpha(D^2 - d^2)$ (See Note 12)	$A = \frac{1}{2} r\ell = 0.00873\, \alpha\, r^2$ $A = \pi r^2 \alpha/360 \text{ deg}$ $A = \frac{1}{2} r^2 \text{ rad } \alpha$ (See Note 13)
Volume (V)				
Height (h)	$h = d - r$			
Sides (a, b or c)	$S = R\theta = \frac{1}{2} D\theta$ $S = D \cos^{-1} d/R$	$b = R - r$ $b = A/2\pi R'$		
Radius (r, r′, R, R′)	$r = R$ $r = C/6.2832$ $r = \sqrt{A/3.1416}$ $r = 0.564\sqrt{A}$	$R' = \frac{1}{2}(R + r)$	$(R^2 - r^2) = \dfrac{A}{0.00873\, \alpha}$ $(D^2 - d^2) = \dfrac{A}{0.00218\, \alpha}$	$r = 2A/\ell$ $r = 57.296\ell/\alpha$
Minor Dia. (d), Major Dia. (D)	$d = D$ $d = C/3.1416$ $d = \sqrt{A/0.7854}$ $d = 1.128\sqrt{A}$ (See Note 9)			
Length of Element (ℓ)	$\ell = 2\sqrt{d^2 - r^2}$ $\ell = 0.01745 r\theta$ $\ell = 2R \sin \theta/2$ $\ell = \dfrac{2A}{r}$ $\ell = 2d \tan \theta/2$ (See Note 10) $\ell = \dfrac{r\theta\pi}{180}$			$\ell = \dfrac{r\, \alpha\, 3.1416}{180°}$ $\ell = 0.01745\, r\, \alpha$ $\ell = 2A/r$
Lateral Area (LA)				

Name of Figure	CIRCULAR SEGMENT	CONE	CUBE	CYCLOID
Classification	PLANE FIGURE	SOLID	SOLID	PLANE FIGURE
Kind of Figure		(Ref. Rt. Circular Cone)		
Area (A)	$A = \frac{1}{2}[r\ell - c(r-h)]$ $A = \frac{1}{2}r^2(\text{rad } \alpha - \sin \alpha)$ $A = \frac{1}{2}[r(\ell - c) + ch]$	$A = 3.1416\,rs$ $A = 3.1416r\sqrt{r^2 + h^2}$ $A = 1.5708\,ds$	Total surface area $= 6\,a^2$	$A = 3\pi r^2 = 9.4248\,r^2$ $A = 2.3562\,d^2$ $A = 3 \times$ area of generating circle
Volume (V)		$V = 1.0472\,r^2 h$ $V = 0.2618\,d^2 h$	$V = a^3$	
Height (h)	$h = r - \frac{1}{2}\sqrt{4r^2 - c^2}$	$h = \dfrac{V}{1.0472\,r^2}$ $h = \dfrac{V}{0.2618\,d^2}$		
Sides (a, b or c)	$c = 2\sqrt{h(2r - h)}$	$\text{side } s = \sqrt{r^2 + h^2}$ $\text{side } s = \sqrt{\dfrac{d^2}{4} + h^2}$	$a = \sqrt[3]{V}$	
Radius (r, r', R, R')	$r = \dfrac{c^2 + 4h^2}{8h}$ $r = \dfrac{57.296\,\ell}{\alpha}$	$r = \dfrac{A}{3.1416\,s}$ $r = \sqrt{\dfrac{V}{1.0472\,h}}$		$r = \dfrac{\ell}{8} = \sqrt{\dfrac{A}{3\pi}}$ $r = \sqrt{\dfrac{A}{9.4248}}$
Minor Dia. (d), Major Dia. (D)				$d = \ell/4$ $d = \sqrt{\dfrac{A}{2.3562}}$
Length of Element (ℓ)	$\ell = 0.01745\,r\,\alpha$ $\ell = \dfrac{\alpha r}{57.296}$		length of diagonal $d = \ell = a\sqrt{3}$	$\ell = 8r$ $\ell = 4d$
Lateral Area (LA)				

22

	CYLINDER	ELLIPSE	ELLIPSOID	EQUILATERAL TRIANGLE
Name of Figure	CYLINDER	ELLIPSE	ELLIPSOID	EQUILATERAL TRIANGLE
Classification	Solid	Plane Figure	Solid	Plane Figure
Kind of Figure				
Area (A)	S = Area of cylindrical surface = 6.2832 rh = 3.1416 dh (See Note 14)	$A = \pi ab = 3.1416\ ab$ Area of shaded segment = $xy + ab\ \sin^{-1}(x/a)$		$A = \frac{1}{4}a^2\sqrt{3}$ $A = 0.250\ a^2\sqrt{3}$ $A = 0.43301\ a^2$
Volume (V)	$V = 3.1416\ r^2h$ $V = 0.7854\ d^2h$ Also, V = (Area of base) × (Altitude)		$V = 4/3\pi\ abc$ (where a, b, c = semiaxes)	
Height (h)	$h = V/3.1416\ r^2$ $h = V/0.7854\ d^2$ $h = S/6.2832\ r$ $h = S/3.1416\ d$			
Sides (a, b or c)		$a = A/3.1416\ b$ $b = A/3.1416\ a$	$a = V/\frac{4}{3}\pi bc$ $b = V/\frac{4}{3}\pi ac$ $c = V/\frac{4}{3}\pi ab$	$a = \sqrt{\dfrac{A}{0.43301}}$
Radius (r, r', R, R')	$r = S/6.2832\ h$ $r = \sqrt{\dfrac{V}{3.1416\ h}}$			
Minor Dia. (d), Major Dia. (D)	$d = \dfrac{S}{3.1416\ h}$ $d = \sqrt{\dfrac{V}{0.7854\ h}}$			
Length of Element (ℓ)		(See Note 15)		
Lateral Area (LA)	$LA = 2\pi rh = Ph$ Here B = area of base and P = perimeter of base.			

23

Name of Figure	FRUSTUM OF CONE	FRUSTUM OF PYRAMID	FRUSTUM OF REGULAR PYRAMID	FRUSTUM OF RIGHT-CIRCULAR CONE
Classification	Solid	Solid	Solid	Solid
Kind of Figure				
Area (A)	A = Area of conical surface = $\pi s(R + r)$ $A = 1.5708\,s(D + d)$	Area of top = A_1 Area of base = A_2		
Volume (V)	$V = 1.0472h(R^2 + Rr + r^2)$ $V = 0.2618h(D^2 + Dd + d^2)$ (See Note 16)	$V = \dfrac{h}{3}(A_1 + A_2 + \sqrt{A_1 \times A_2})$	$V = \frac{1}{6}\,hran[1 + (a'/a) + (a'/a)^2]$ (See Note 17)	$V = \frac{1}{3}\pi h(r^2 + rr' + r'^2)$ $V = \frac{1}{3}\pi r^2 h[1 + (r'/r) + (r'/r)^2]$ $V = \frac{1}{4}\pi h[(r + r')^2 + \frac{1}{3}(r - r')^2]$
Height (h)	$h = \dfrac{V}{1.0472(R^2 + Rr + r^2)}$ $h = \dfrac{V}{0.2618(D^2 + Dd + d^2)}$	$h = \dfrac{3V}{(A_1 + A_2 + \sqrt{A_1 \times A_2})}$	$h = \dfrac{6V}{ran\left[1 + (a'/a) + \left(\dfrac{a'}{a}\right)^2\right]}$	$h = \dfrac{3V}{\pi r^2[1 + (r'/r) + (r'/r)^2]}$ $h = \dfrac{3V}{\pi(r^2 + rr' + r'^2)}$ $h = \dfrac{4V}{\pi[(r + r')^2 + \frac{1}{3}(r - r')^2]}$
Sides (a, b or c)	$a = R - r$			
Radius (r, r', R, R')	$r = -a + R$			
Minor Dia. (d), Major Dia. (D)				
Length of Element (ℓ)	length of side $s = a^2 + h^2$ $s = \sqrt{(R - r)^2 + h^2}$			
Lateral Area (LA)	$LA = \frac{1}{2}$(sum of perimeters of bases) × (slant height)	$LA = \frac{1}{2}$(sum of perimeters of base) × (slant height)	$LA = \frac{1}{2}sn(r + r')$, where $s = \sqrt{(r - r')^2 + h^2}$	$LA = \pi s(r + r')$ where $s = \sqrt{(r - r')^2 + h^2}$

Name of Figure	HOLLOW CYLINDER (RIGHT AND CIRCULAR)	HOLLOW SPHERE OR SPHERICAL SHELL	HYPERBOLA	OBELISK (FRUSTUM OF A RECTANGLE PYRAMID)
Classification	Solid	Solid	Plane Figure	Solid
Kind of Figure				
Area (A)			$A = \text{area } BCD$ $A = \dfrac{xy}{2} - \dfrac{ab}{2}$ hyp. log $(x/a + y/b)$ (See Note 18)	
Volume (V)	$V = \pi h(R^2 - r^2) = \pi h t(D - t)$ $V = \pi h t(2R - t) = \pi h t(d + t)$ $V = \pi h t(2r + t) = \pi h t(R + r)$ $V = 0.7854 h(D^2 - d^2)$	$V = 4.1888(R^3 - r^3)$ $V = 0.5236(D^3 - d^3)$ $V = 4\pi R_1^2 t + \frac{1}{3}\pi t^3$		$V = \frac{1}{6}h[(2a + a_1)b + (2a_1 + a)b_1]$ $V = \frac{1}{6}h[ab + (a + a_1)$ $(b + b_1) + a_1 b_1]$
Height (h)	$h = V/\pi(R^2 - r^2) = V/\pi t(d + t)$ $h = V/\pi t(2R - t) = V/\pi t(d + t)$ $h = V/\pi t(2r + t) = V/\pi t(R + r)$ $h = V/0.7854(D^2 - d^2)$			$h = \dfrac{6V}{[(2a + a_1)b + (2a_1 + a)b_1]}$ $h = \dfrac{6V}{[ab + (a + a_1)(b + b_1) + a_1 b_1]}$
Sides (a, b or c)	$t = R - r$	$t = b = \text{thickness}$ $= R - r$	a and b are half-axes x and y are coordinates of point P	
Radius (r, r′, R, R′)	$r = R - b$ $R = r + b$	$r = R - b$ $R = r + b$ $R_1 = \text{mean radius}$ $R_1 = \frac{1}{2}(R + r)$		
Minor Dia. (d), Major Dia. (D)	$D' = \text{mean diameter}$ $D' = \frac{1}{2}(d + D)$ $D' = D - b = d + b$			
Length of Element (ℓ)				
Lateral Area (LA)				

Name of Figure	OBTUSE-ANGLE TRIANGLE	PARABOLA: CASE I	PARABOLA: CASE II	PARABOLA: CASE III
Classification	Plane Figure	Plane Figure	Plane Figure	Plane Figure
Kind of Figure				
Area (A)	$A = bh/2$ $A = \dfrac{b}{2}\sqrt{a^2 - \left(\dfrac{c^2 - a^2 - b^2}{2b}\right)^2}$ (See Note 19)	$A = 2\ell d/3$		$A = \dfrac{2}{3}xy$ (See Note 21)
Volume (V)				
Height (h)	$h = 2A/b$	$d_1 =$ Perpendicular distance from base line at point P to the parameter line height of d_1 $= \dfrac{d}{\ell^2}(\ell^2 - \ell_1^2)$		$y = \dfrac{A}{2/3\,x}$ where $y =$ height
Sides (a, b or c)	$b = 2A/h$	Width of ℓ_1 $= \ell\sqrt{\dfrac{d - d_1}{d}}$	$\dfrac{P}{2} =$ Half the distance from focus to directrix	
Radius (r, r', R')				
Minor Dia. (d), Major Dia. (D)				
Length of Element (ℓ)		$\ell_1 =$ Width of the curve length of arc $=$ $\ell\left[1 + \dfrac{2}{3}\left(\dfrac{2d}{\ell}\right)^2 - \dfrac{2}{5}\left(\dfrac{2d}{\ell}\right)^4 + \ldots\right]$	(See Note 20)	$x = \dfrac{A}{2/3\,y}$ where $x =$ base
Lateral Area (LA)				

Name of Figure	PARABOLA: CASE IV	PARABOLA: CASE V	PARABOLOID	PARABOLOID OF REVOLUTION
Classification	Plane Figure	Plane Figure	Solid	Solid
Kind of Figure				
Area (A)	$A = \frac{2}{3}\,ch$ (See Note 22)		$A = \frac{2\pi}{3p}\left[\sqrt{\left(\frac{d^2}{4} + p^2\right)^3} - p^3\right]$ in which $p = d^2/8h$	
Volume (V)			$V = \frac{1}{2}\,\pi r^2 h$ $V = 0.3927\ d^2 h$	$V = \frac{1}{2}\,\pi r^2 h$ $V = \frac{1}{2}$ volume of circumscribed cylinder.
Height (h)	$h = \dfrac{A}{\frac{2}{3}\,c}$		$h = \dfrac{V}{\frac{1}{2}\,\pi r^2}$ $h = \dfrac{V}{0.3927\ d^2}$	$h = \dfrac{V}{\frac{1}{2}\,\pi r^2}$
Sides (a, b or c)	$\text{side } c = \dfrac{A}{\frac{2}{3}\,h}$			
Radius (r, r', R, R')			$r^2 = \dfrac{V}{\frac{1}{2}\,\pi h}$ or $r = \sqrt{\dfrac{V}{\frac{1}{2}\,\pi h}}$	$r^2 = \dfrac{V}{\frac{1}{2}\,\pi h}$ $r = \sqrt{\dfrac{V}{\frac{1}{2}\,\pi h}}$
Minor Dia. (d), Major Dia. (D)			$d = \sqrt{\dfrac{V}{0.3927\ h}}$	
Length of Element (ℓ)		length of arc OP = s $s = \frac{1}{2}\,PT + \frac{1}{2}\,p\ell n \cot \frac{1}{2}\,\mu$ (See Note 23)		
Lateral Area (LA)				

Name of Figure	PARABOLOIDAL SEGMENT	PARALLELOGRAM: CASE I	PARALLELOGRAM: CASE II	PORTION OF CYLINDER: CASE I
Classification	Solid	Plane Figure	Plane Figure	Solid
Kind of Figure				
Area (A)		$A = bh = ab \sin C$ $A = \frac{1}{2} D_1 D_2 \sin \mu$ (See Note 24)	$A = ab$ (See Note 25)	
Volume (V)	$V = 1.5708h\ (R^2 + r^2)$ $V = 0.3927h\ (D^2 + d^2)$			$V = 1.5708r^2(h_1 + h_2)$ $V = 0.3927d^2(h_1 + h_2)$
Height (h)	$h = \dfrac{V}{1.5708\ (R^2 + r^2)}$ $h = \dfrac{V}{0.3927\ (D^2 + d^2)}$	$h = A/b$	Here $a = h$ or $a = A/b$	
Sides (a, b or c)		$b = A/h$	$a = A/b$ $b = A/a$	
Radius (r, r′, R, R′)				$r = \sqrt{\dfrac{V}{1.5708\ (h_1 + h_2)}}$ $r = \dfrac{S}{3.1416\ (h_1 + h_2)}$
Minor Dia. (d), Major Dia. (D)				$d = \sqrt{\dfrac{V}{0.3927\ (h_1 + h_2)}}$ $d = \dfrac{S}{1.5708\ (h_1 + h_2)}$
Length of Element (ℓ)				
Lateral Area (LA)				$S = $ area of cyl. surface $S = \pi r(h_1 + h_2)$ $S = 1.5708d(h_1 + h_2)$

Name of Figure	PORTION OF CYLINDER: CASE II	PRISM: CASE I	PRISM: CASE II	PRISMOIDAL FORMULA: CASE I
Classification	Solid	Solid	Solid	Solid
Kind of Figure			For any prism or cylinder, see equations below.	
Area (A)	S = area of cylindrical surface = (ad $\pm$ b $\times$ length of arc ABC) h/r $\pm$ b. (See Note 26)	A = area of end surface $A = \dfrac{V}{h}$ (See Note 27)	Area of base = Volume/Altitude $= \dfrac{V}{h}$	(See Note 28)
Volume (V)	$V = \left(\dfrac{2}{3} a^3 \pm b \times \text{area ABC}\right) h/r \pm b$ (See Note 26)	$V = A \times h$	V = Ah or Volume = (area of base) $\times$ (altitude)	$V = \dfrac{1}{6} h(A_1 + A_2 + 4A_3)$ (See Note 28)
Height (h)		$h = \dfrac{V}{A}$	$h = \dfrac{V}{A}$	$h = \dfrac{6V}{(A_1 + A_2 + 4A_3)}$
Sides (a, b or c)				
Radius (r, r', R, R')				
Minor Dia. (d), Major Dia. (D)				
Length of Element (ℓ)				
Lateral Area (LA)			LA = (Perimeter of right section) $\times$ (lateral edge)	

	PRISMOIDAL FORMULA: CASE II	PYRAMID	QUADRILATERAL INSCRIBED IN A CIRCLE	RECTANGLE: CASE I
Classification	SOLID	SOLID	PLANE FIGURE	PLANE FIGURE
Kind of Figure				
Area (A)	(See Note 29)	B = Base area (See Note 30)	$A = \frac{1}{2} D_1 D_2 \sin \mu$ $A = \sqrt{(s-a)(s-b)(s-c)(s-d)}$ $A = \frac{1}{2}(ac + bd) \sin \mu$	$A = ab$ $A = a\sqrt{d^2 - a^2}$ $A = b\sqrt{d^2 - b^2}$
Volume (V)	To find the volume of any prism, pyramid or frustum of a pyramid, see Note 28 and formulas below: $V = h/6(A_1 + 4A_m + A_2)$	$V = 1/3\ Bh$ $V = nsrh/6$ $V = \dfrac{nsh}{6}\sqrt{R^2 - \dfrac{s^2}{4}}$		
Height (h)	$h = \dfrac{6V}{(A_1 + 4A_m + A_2)}$	$h = \dfrac{V}{1/3\ B}$ $h = \dfrac{6V}{nsr}$		
Sides (a, b or c)			$s = \frac{1}{2}(a + b + c + d)$	$a = A/b = \sqrt{d^2 - b^2}$ $b = A/a = \sqrt{d^2 - a^2}$
Radius (r, r', R, R')		$r = \dfrac{6V}{nsh}$		
Minor Dia. (d), Major Dia. (D)			$D_1 = \dfrac{A}{\frac{1}{2} D_2 \sin \mu}$ $D_2 = \dfrac{A}{\frac{1}{2} D_1 \sin \mu}$	
Length of Element (ℓ)		Let $s = \ell$ $\ell = \dfrac{6V}{nrh}$		Let diagonal $d = \ell$ $\ell = \sqrt{a^2 + b^2}$
Lateral Area (LA)				

Name of Figure	RECTANGLE: CASE II	RECTANGULAR PARALLELOPIPED	REGULAR HEXAGON	REGULAR OCTAGON
Classification	Plane Figure	Solid	Plane Figure	Plane Figure
Kind of Figure		A Rectangular Parallelopiped is a six-sided prism whose faces are parallelograms		
Area (A)	$a = ab$ $a = \frac{1}{2} D^2 \sin \mu$ where μ = angle between diagonals DD	Total surface area = $2(ab + bc + ca)$, where a, b, and c are the lengths of the sides	$A = 2.598\ s^2$ $A = 2.598\ R^2$ $A = 3.464\ r^2$	$A = 4.828\ s^2$ $A = 2.828\ R^2$ $A = 3.314\ r^2$
Volume (V)		$V = abc$		
Height (h)				
Sides (a, b or c)		$a = V/bc$ $b = V/ac$ $c = V/ab$	side $s = 1.155\ r = R$ $= \sqrt{\dfrac{A}{2.598}}$	side $s = 0.765\ R$ side $s = 0.828\ r$ side $s = \sqrt{\dfrac{A}{4.828}}$
Radius (r, r', R, R')			$R = s = 1.155r = \sqrt{A/2.598}$ $r = R/1.155 = s/1.155$ $r = 0.866s = 0.866\ R$ $r = \sqrt{A/3.464}$	$R = 1.307s = 1.082r$ $R = s/0.765 = \sqrt{A/2.828}$ $r = 1.207s = 0.924\ R$ $r = R/1.082 = \sqrt{A/3.314}$
Minor Dia. (d), Major Dia. (D)				
Length of Element (ℓ)	Let Diagonal $D^2 = \ell^2$ $\ell^2 = \dfrac{A}{\frac{1}{2} \sin \mu}$	$d = \ell$ = length of diagonal $\ell = \sqrt{a^2 + b^2 + c^2}$		
Lateral Area (LA)				

Name of Figure	REGULAR POLYGON	REGULAR POLYGON OF n SIDES	REGULAR PRISM	REGULAR PYRAMID
Classification	Plane Figure	Plane Figure	Solid	Solid
Kind of Figure				
Area (A)	$A = nsr/2$ $= \dfrac{ns}{2}\sqrt{R^2 - \dfrac{s^2}{4}}$ (See Note 31)	$A = \frac{1}{4}\, n\, a^2\, \text{ctn}\, \dfrac{180°}{n}$ (See Note 32)	$B = $ area of base	(See Note 33)
Volume (V)			$V = \frac{1}{2}\, nrah$ $V = Bh$, where $n = $ number of sides	$V = \frac{1}{3}$ altitude $\times$ (area of base) $V = \frac{1}{6}\, hran$
Height (h)			$h = \dfrac{V}{\frac{1}{2}\, nra} = \dfrac{V}{B}$ $h = \dfrac{LA}{na} = \dfrac{LA}{P}$	$h = \dfrac{V}{\frac{1}{6}\, ran}$
Sides (a, b or c)	side $s = \dfrac{2A}{nr}$ side $s = 2\sqrt{R^2 - r^2}$	side $a = 2r \tan \alpha/2$ side $a = 2R \sin \alpha/2$	side $a = $ $\dfrac{V}{\frac{1}{2}\, nrh}$ $= \dfrac{LA}{nh}$	side $a = \dfrac{V}{1/6\ hrn}$ side $a = LA/\frac{1}{2}\, sn$ $s = \sqrt{r^2 + h^2}$
Radius (r, r', R, R')	$R = \sqrt{r^2 + \dfrac{s^2}{4}}$ $r = \sqrt{R^2 - \dfrac{s^2}{4}}$	$R = a/2 \csc \dfrac{180°}{n}$ $r = a/2\ \text{ctn}\ \dfrac{180°}{n}$	$r = \dfrac{V}{\frac{1}{2}\, nah}$	$r = \dfrac{V}{\frac{1}{6}\, han}$
Minor Dia. (d), Major Dia. (D)				
Length of Element (ℓ)				
Lateral Area (LA)			$LA = nah$ $LA = Ph$, where $P = $ Perimeter of base	$LA = \frac{1}{2}$ Slant height $\times$ Perimeter of base $LA = \frac{1}{2}\, san$

Name of Figure	RIGHT-ANGLE TRIANGLE	RIGHT-CIRCULAR CONE	RIGHT-CIRCULAR CYLINDER	SEGMENT OF PARABOLA
Classification	Plane Figure	Solid	Solid	Plane Figure
Kind of Figure				
Area (A)	$A = \frac{1}{2} ab = \frac{1}{2} a^2 \cot A$ $A = \frac{1}{2} b^2 \tan A$ $A = \frac{1}{4} c^2 \sin 2A$		B = area of base	Area of GCE = $A = 2/3$ area of parallelogram GEDB, area of segment GCD = 2/3 GE (CF)
Volume (V)		$V = \frac{1}{3} \pi r^2 h$	$V = \pi r^2 h$ $V = Bh$	
Height (h)		$h = \dfrac{V}{\frac{1}{3}\pi r^2}$	$h = \dfrac{V}{\pi r^2} = \dfrac{V}{B}$ $h = \dfrac{LA}{2\pi r} = \dfrac{LA}{P}$	
Sides (a, b or c)	$a^2 + b^2 = c^2$ $a = \sqrt{b^2 - c^2}$ $b = \sqrt{a^2 - c^2}$ $c = \sqrt{a^2 + b^2}$	side s = slant height side $s = \sqrt{r^2 + h^2}$ side $s = \dfrac{LA}{\pi r}$		Line passing thru P and then center of directrix defines points G, F and E. Line containing points B, C and D is then drawn parallel to line G, F and E. Horizontal lines BG and DE are parallel to TX.
Radius (r, r', R, R')		$r = \dfrac{LA}{\pi s}$ $r = \sqrt{\dfrac{V}{1/3\ \pi h}}$	$r = \dfrac{LA}{2\pi h}$ $r = \sqrt{\dfrac{V}{\pi h}}$	
Minor Dia. (d), Major Dia. (D)				
Length of Element (ℓ)				
Lateral Area (LA)		$LA = \pi rs$	$LA = 2\pi rh = Ph$ where P = Perimeter of base	

33

Name of Figure	SEGMENT OF PARABOLOID OF REVOLUTION	SPANDREL OR FILLET	SPECIAL UNGULA OF A RIGHT-CIRCULAR CYLINDER	SPHERE
Classification	Solid	Plane Figure	Solid	Solid
Kind of Figure				
Area (A)		$A = r^2 - \dfrac{\pi r^2}{4}$ $A = 0.215\ r^2$ $A = 0.1075\ c^2$	Note: Upper surface is a semiellipse	$A = 4\pi r^2 = \pi d^2$ $A = 12.5664\ r^2$ $A = 3.1416\ d^2$
Volume (V)	Volume of segment $= \frac{1}{2}\ \pi(R^2 + r^2)h$		$V = \frac{2}{3}\ r^2\ h$	$V = \dfrac{4\pi r^3}{3} = \dfrac{\pi d^3}{6}$ $V = 4.1888\ r^3$ $V = 0.5236\ d^3$
Height (h)	$h = \dfrac{V}{\frac{1}{2}\ \pi(R^2 + r^2)}$		$h = \dfrac{V}{2/3\ r^2}$ $h = \dfrac{LA}{2r}$	
Sides (a, b or c)		$c = \sqrt{\dfrac{A}{0.1075}}$ where c = chord		
Radius (r, r', R, R')	$(R^2 + r^2) = \dfrac{V}{\frac{1}{2}\ \pi h}$	$r = \sqrt{\dfrac{A}{0.215}}$	$r = \dfrac{LA}{2h}$ $r = \sqrt{\dfrac{V}{2/3\ h}}$	$r = \sqrt[3]{\dfrac{3V}{4\pi}}$ $r = 0.6204\ \sqrt[3]{V}$
Minor Dia. (d), Major Dia. (D)				$d = \sqrt{\dfrac{A}{3.1416}}$ $d = \sqrt[3]{\dfrac{V}{0.5236}}$
Length of Element (ℓ)				
Lateral Area (LA)			$LA = 2rh$	

Name of Figure	SPHERICAL SECTOR	SPHERICAL SEGMENT	SPHERICAL SEGMENT OF THE BASE ZONE (SPHERICAL CAP)	SPHEROID (OR ELLIPSOID OF REVOLUTION)
Classification	Solid	Solid	Solid	Solid
Kind of Figure				Prismoidal Formula Applicable (See Note 34)
Area (A)	$A = \pi r (2h + \frac{1}{2} c)$	$A = 2\pi rh$ $A = 6.2832\ rh$ $A = 3.1416\left(\frac{c^2}{4} + h^2\right)$ $A = $ Area of spherical surface		A and B = areas of bases. M = area of a plane section midway between the bases.
Volume (V)	$V = \dfrac{2\pi r^2 h}{3}$ $V = 2.0944\ r^2 h$	$V = \pi h^2 (r - h/3)$ $V = \pi h \left(\dfrac{c^2}{8} + \dfrac{h^2}{6}\right)$	$V = \frac{1}{6}\ \pi h(3a^2 + h^2)$ $V = 1/3\ \pi h^2(3r - h)$	$V = \frac{1}{6}\ h(A + B + 4M)$
Height (h)	$h = \dfrac{3V}{2\pi r^2}$ $h = \dfrac{V}{2.0944\ r^2}$	$h = \dfrac{A}{2\pi r}$ $h = \dfrac{A}{6.2832\ r}$	$h = \dfrac{LA}{2\pi r}$	$h = \dfrac{6V}{(A + B + 4M)}$
Sides (a, b or c)	$c = 2\sqrt{h(2r - h)}$ where c = chord	$c = 2\sqrt{h(2r - h)}$ where c = chord		
Radius (r, r′, R, R′)	$r = \sqrt{\dfrac{3V}{2\pi h}}$ $r = \sqrt{\dfrac{V}{2.0944\ h}}$	$r = \dfrac{c^2 + 4h^2}{8h}$ $r = \dfrac{A}{6.2832\ h}$	$r = \dfrac{LA}{2\pi h}$	
Minor Dia. (d), Major Dia. (D)				
Length of Element (ℓ)				
Lateral Area (LA)			LA = Lateral axis LA (of zone) $LA = 2\pi rh$ $LA = \pi(a^2 + h^2)$	

35

Name of Figure	SPHERICAL WEDGE	SPHERICAL ZONE	SQUARE	SQUARE PRISM
Classification	Solid	Solid	Plane Figure	Solid
Kind of Figure				
Area (A)	$A = \dfrac{\alpha}{360°} \times (4\pi r^2)$ $A = 0.0349 \alpha r^2$ $A =$ Area of spherical surface	$A = 2\pi rh$ $A = 6.2832\ rh$ $A =$ Area of spherical surface	$A = s^2$ $A = \frac{1}{2}\ d^2$	
Volume (V)	$V = \dfrac{\alpha}{360°}\left(\dfrac{4\pi r^3}{3}\right)$ $V = 0.0116 \alpha r^3$	$V = 0.5236\ h \times$ $\left(\dfrac{3c_1^2}{4} + \dfrac{3c_2^2}{4} + h^2\right)$		$V = abc$
Height (h)		$h = \dfrac{A}{2\pi r}$ $h = \dfrac{A}{6.2832\ r}$		
Sides (a, b or c)			side $s = \sqrt{A}$ side $s = 0.7071\ d$ side $s = d/1.414$	$a = V/bc$ $b = V/ac$ $c = V/ab$
Radius (r, r′, R, R′)	$r = \sqrt[3]{\dfrac{V}{0.0116\ \alpha}}$ $r = \sqrt{\dfrac{A}{0.0349\ \alpha}}$	$r = \dfrac{A}{2\pi h}$ $r = \dfrac{A}{6.2832\ h}$		
Minor Dia. (d), Major Dia. (D)				
Length of Element (ℓ)			$\ell = d = 1.414\ s$ $\ell = 1.414\sqrt{A}$	
Lateral Area (LA)				

36

Name of Figure	TORUS	TRAPEZIUM	TRAPEZOID	WEDGE
Classification	Solid	Plane Figure	Plane Figure	Solid
Kind of Figure				
Area (A)	$A = 4\pi^2 Rr = 19.739\ Rr^2$ $A = \pi^2 Dd$ $A = 9.8696\ Dd$	$A = \dfrac{(H + h)a + bh + cH}{2}$	$A = \frac{1}{2} h(a + b)$ $A = \frac{1}{2} D_1 D_2 \sin \mu$	
Volume (V)	$V = 2\pi^2\ Rr^2$ $V = 19.739\ Rr^2$ $V = \pi^2/4\ Dd^2$ $V = 2.4674\ Dd^2$			$V = \dfrac{(2a + c)\ bh}{6}$
Height (h)			$h = \dfrac{A}{\frac{1}{2} (a + b)}$	$h = \dfrac{6V}{(2a + c)\ b}$
Sides (a, b or c)			Diagonal $D_1 =$ $\dfrac{A}{\frac{1}{2} D_2 \sin \mu}$ Diagonal $D_2 =$ $A/\frac{1}{2} D_1 \sin \mu$	$b = \dfrac{6V}{(2a + c)\ h}$
Radius (r, r′, R, R′)	$r = \sqrt{\dfrac{V}{2\pi^2 R}} = \sqrt{\dfrac{V}{19.739\ R}}$ $R = \dfrac{V}{2\pi^2 R} = \dfrac{V}{19.739\ r^2}$ (See Note 35)			
Minor Dia. (d), Major Dia. (D)				
Length of Element (ℓ)				
Lateral Area (LA)				

Notes

1. In an acute-angle triangle, if

$$s' = \frac{1}{2}(a + b + c) \text{ then:}$$
$$A = \sqrt{s'(s' - a)(s' - b)(s' - c)}$$

2. These equations are applicable to either an anchor ring or torus.

3. In a pyramid or conical frustum:

   B = Area of lower base
   B'= Area of upper base
   P = Perimeter of lower base
   P'= Perimeter of upper base

4. In any spherical segment (zone), if the inscribed frustum of a cone is removed from the spherical segment, the volume remaining is:

$$V = \frac{1}{6}\pi hc^2 \text{ where:}$$
   c = Slant height of frustum = $\sqrt{h^2 + (a - a_1)^2}$

5. The area in any triangle is equal to the following equations:

$$A = \frac{4}{3}\sqrt{t(t - m_1)(t - m_2)(t - m_3)}$$
$$A = r^2 \cot \frac{1}{2}A \cot \frac{1}{2}B \cot \frac{1}{2}C$$
$$A = 2R^2 \sin A \sin B \sin C$$
$$A = \pm\frac{1}{2}[(x_1y_2 - x_2y_1) + (x_2y_2 - x_3y_2) + (x_3y_1 - x_1y_3)]$$

   where (x_1y_1), (x_2y_2) and (x_3y_3) are the coordinates of vertices.

6. For a truncated triangular prism with lateral edges a, b and c:

$$\ell = K = \frac{1}{3}(a + b + c) \text{ where:}$$
   K = Distance between centers of gravity of perimeters of the two edges
   N = Area of normal section
   Q = Perimeter of normal section

7. This equation is applicable if the sides of the barrel are bent to the arc of a circle.

8. This equation is applicable if the sides of the barrel are bent to the arc of a parabola.

9. In a circle the diameter equals:

$$d = \frac{1}{2}\sqrt{4d^2 - R^2}$$
$$d = R\left(\frac{1}{2}\cos\theta\right)$$
$$d = \frac{\ell}{2}\left(\frac{\ell \operatorname{ctn}\theta}{2}\right)$$

10. In a circle the following equations are applicable:

$$\theta = \frac{S}{R}$$
$$\theta = \frac{2S}{D}$$
$$\theta = 2\cos^{-1}\frac{d}{R}$$
$$\theta = 2\tan^{-1}\frac{\ell}{2d}$$
$$\theta = 2\sin^{-1}\frac{\ell}{D}$$
$$S = R\theta$$
$$S = \frac{1}{2}D\theta$$
$$S = D\cos^{-1}\frac{d}{R} \text{ where:}$$
   S = Length of arc subtended by θ
$$\ell = 2\sqrt{R^2 - d^2}$$
$$\ell = 2R\left(\frac{1}{2}\sin\theta\right)$$
$$\ell = 2d\left(\frac{1}{2}\tan\theta\right)$$
$$d = \frac{1}{2}\sqrt{4R^2 - \ell^2}$$
$$d = \frac{R\cos\theta}{2}$$
$$d = \frac{1}{2}\ell\left(\frac{1}{2}\operatorname{ctn}\theta\right)$$
$$R = \frac{S}{\theta}$$
$$R = \frac{\ell}{2\left(\frac{1}{2}\sin\theta\right)}$$
$$D = \frac{2S}{\theta}$$
$$D = \frac{S}{\cos^{-1}\frac{d}{R}}$$

Area of circumscribed polygon:

$$A = nR^2 \tan\frac{\pi}{n}$$
$$A \text{ (circle)} = \pi R^2$$
$$A \text{ (circle)} = \frac{1}{4}\pi D^2$$
$$A \text{ (sector)} = \frac{1}{2}RS$$
$$A \text{ (sector)} = \frac{1}{2}R^2\theta$$
$$A \text{ (segment)} = A(\text{sector}) - A(\text{triangle})$$
$$A \text{ (segment)} = \frac{1}{2}R^2(\theta - \sin\theta)$$
$$A \text{ (segment)} = R^2\cos^{-1}\frac{(R - h)}{R} - (R - h)\sqrt{2Rh - h^2}$$

Notes

Area of inscribed polygon:

$$A = \frac{1}{2} nR^2 \sin \frac{2\pi}{n}$$

The perimeter of a n − sided regular polygon circumscribed about a circle is:

$$P = 2nR \tan \frac{\pi}{n}$$

The perimeter of a n − sided regular polygon inscribed in a circle is:

$$P = 2nR \sin \frac{\pi}{n}$$

The radius of a circle inscribed in a triangle of sides a, b and c is:

$$r = \sqrt{\frac{(s-a)(s-b)(s-c)}{s}} \text{ where:}$$

$$s = \frac{1}{2}(a + b + c)$$

The radius of a circle circumscribed about a triangle is:

$$R = \frac{abc}{4\sqrt{s(s-a)(s-b)(s-c)}}$$

11. In a circular ring:

$$R' = \frac{1}{2}(R + r)$$
$$b = R - r$$
$$R' = \text{Mean radius}$$

12. In a circular ring sector:

$$\alpha = \frac{A}{0.00873(R^2 - r^2)}$$
$$\alpha = \frac{A}{0.00218(D^2 - d^2)}$$

13. In a circular sector:

$$\alpha = 57.296 \frac{\ell}{r}$$
$$\alpha = \frac{\ell}{0.01745r}$$
$$\ell = r \text{ rad } \alpha \text{ where:}$$
$$\alpha = \text{Radian measure of angle } \alpha$$
$$\ell = \text{Length of arc}$$

14. The total area of a cylindrical surface is:

$$A = 3.1416d \left(\frac{1}{2}d + h \right)$$
$$A = 6.2832r(r + h)$$

15. The length of the perimeter of an ellipse is:

$$\ell = \pi(a + b) K \text{ where :}$$

$$K = (1 + \frac{1}{4}m^2 + \frac{1}{64}m^4 + \frac{1}{256}m^6 + \ldots \ldots)$$

The length of the perimeter (approximate) is:
$$\ell = \pi \sqrt{2(a^2 + b^2)}$$

The length of the perimeter (closer approximation) is:

$$\ell = \pi \sqrt{2(a^2 + b^2) - \frac{(a-b)^2}{2.20}}$$

16. The volume of a frustum of a cone is:

$$V = \frac{1}{3}(A_1 + A_2 + \sqrt{(A_1)(A_2)}h \text{ where:}$$
$$h = \text{Altitude}$$
$$A_1 \text{ and } A_2 = \text{Areas of the bases}$$

17. In a frustum of regular pyramid:

$$LA = (\text{Slant height})(\text{perimeter of mid-section})$$

$$LA = \frac{1}{2}sn(r + r') \text{ where:}$$
$$r, r' = \text{Radii of the inscribed circles}$$

18. In an equilateral hyperbola (a = b), the area is:

$$A = a^2 \sinh^{-1} \left(\frac{y'}{a} \right)$$
$$A = a^2 \cosh^{-1} \left(\frac{x}{a} \right)$$

In any hyperbola the shaded area is:

$$A = ab\ell n \left(\frac{x}{a} + \frac{y}{b} \right)$$

19. In an obtuse-angle triangle the area is:

$$A = \sqrt{s(s-a)(s-b)(s-c)} \text{ where:}$$
$$s = \frac{1}{2}(a + b + c)$$

20. The length of an arc in a parabola: case II is:

$$\ell = \frac{p}{2} \left[\sqrt{\frac{2x}{p} \left(1 + \frac{2x}{p} \right)} \right.$$
$$\left. + \text{hyp. log} \left(\sqrt{\frac{2x}{p}} + \sqrt{1 + \frac{2x}{p}} \right) \right]$$

When x is small in proportion to y:

$$\ell = y \left[1 + \frac{2}{3} \left(\frac{x}{y} \right)^2 - \frac{2}{5} \left(\frac{x}{y} \right)^4 \right]$$
$$\ell = \sqrt{y^2 + \frac{4}{3}x^2}$$

21. In a parabola: case III, the area is:

$$A = \frac{2}{3}xy \text{ where:}$$
$$x = \text{Base of rectangle}$$
$$y = \text{Height of rectangle}$$

22. In a parabola: case IV, the shaded area is:

$$A = \frac{2}{3} ch \text{ where:}$$
$$c = \text{Base of rectangle}$$
$$h = \text{Height of rectangle}$$

Notes

23. In a regular parabola: case V:

     c = Any chord drawn from the vertex 0 to point P

     PT = Tangent at P

     Note: OM = OT = X

24. In the parallelogram: case 1:

     $D_1{}^2 + D_2{}^2 = 2(a^2 + b^2)$ where:

     μ = Angle between diagonals D_1 and D_2

25. In the parallelogram: case II, dimension a is measured at right angles to line b.

26. In the portion of cylinder: case II, use the plus sign (+) when the base area is larger and the minus sign (−) when the area is less than $\frac{1}{2}$ the base circle.

27. In a prism: case 1, the Area A of the end surface is found by the formulas for areas of plane figures on the preceding pages. Height h must be measured perpendicular to the end surfaces.

28. In the prismoidal formula: case 1:

     A_1 = Area at one end of the body

     A_2 = Area at the other end of the body

     A_3 = Area at mid-section between two end surfaces

     h = Height of total body

29. In the prismoidal formula: case II:

     A_1 = Area at one end of the body

     A_2 = Area at the other end of the body

     A_m = Area at mid-section parallel to the bases

     h = Altitude

30. In a regular pyramid:

     n = Number of sides

     s = Length of a side

     r = Radius of inscribed circle

     R = Radius of circumscribed circle

     h = Perpendicular distance from vertex to plane in which base lies

31. In a regular polygon:

     $\alpha = \dfrac{360°}{n}$

     $\beta = 180° - \alpha$

     $\gamma = \dfrac{360°}{n}$

32. In a regular polygon of n sides:

     $\alpha = \dfrac{360°}{n}$

     $\alpha = \dfrac{2\pi}{n}$

     $\beta = \left(n - \dfrac{2}{n}\right) 180°$

     $\beta = \left(n - \dfrac{2}{n}\right) \pi$

33. In a regular pyramid:

     $s = \sqrt{r^2 + h^2}$ where:

     r = Radius of inscribed circle

     a = Side

     n = Number of sides

     Note: the vertex of the pyramid is directly above the center of the base

34. In a spheroid or ellipsoid of revolution, the volume of any solid segment made by two planes parallel and perpendicular to the axis of revolution is found by the prismoidal formula:

     $V = \dfrac{1}{6}h(A + B + 4M)$

35. In a torus:

     $r = \dfrac{A}{4\pi^2 R}$

     $r = \dfrac{A}{39.478R}$

     $R = \dfrac{A}{4\pi^2 r}$

     $R = \dfrac{A}{39.478r}$

     $d = \sqrt{\dfrac{V}{\frac{\pi^2}{4}D}}$

     $d = \sqrt{\dfrac{V}{2.4674D}}$

     $d = \dfrac{A}{\pi^2 D}$

     $d = \dfrac{A}{9.8696d}$

     $D = \dfrac{V}{\frac{\pi^2}{4}d^2}$

     $D = \dfrac{V}{2.4674d^2}$

     $D = \dfrac{A}{\pi^2 d}$

     $D = \dfrac{A}{9.8696d}$

Basic laws of algebra

Introduction.

The basic laws of algebra are presented in this article with examples to illustrate their use.

Addition.

Addition is the collecting of several similar quantities into one term or sum and the connecting of dissimilar quantities by their respective signs.

Case 1: When quantities are similar and have the same signs.

Example: $+9axy + 3axy + 7axy + axy = +20axy$

Case 2: When quantities are similar and have different signs.

Example: $+3x + 4y - 2x + 3y = +x + 7y$

Subtraction.

Subtraction of monomials is indicated by placing the $-$ sign between the quantity to be subtracted and that from which it is to be taken.

Example: From $+4a + 3b - 2c + 8d$ subtract
$$+a + 2b + c + 5d =$$
$$+4a + 3b - 2c + 8d - a - 2b - c - 5d =$$
$$3a + b - 3c + 3d$$

Multiplication.

Multiplication is usually divided into three cases:

Case 1: When both multiplicand and multiplier are simple quantities.

Example: $(5ax)(4axy) = 20a^2x^2y$

Case 2: When the multiplicand is a compound and the multiplier is a simple quantity.

Example: $(a^2 + ab + b^2)(4a) = 4a^3 + 4a^2b + 4ab^2$

Case 3: When both multiplicand and multiplier are compound quantities.

Example: $(ab + cd)(ab - cd) = a^2b^2 - c^2d^2$

Division.

The object of algebraic division is to discover one of the factors of a given product, the other factor being given. Division is usually divided into three cases:

Case 1: When both dividend and divisor are monomials.

Example: $\dfrac{48a^3b^3c^2d}{12ab^2c} = 4a^2bcd$

Case 2: When the dividend is a polynomial and the divisor a monomial.

Example: $\dfrac{6a^2x^4y^6 - 12a^3x^3y^6 + 15a^4x^5y^3}{+3a^2x^2y^2} =$
$$+2x^2y^4 - 4\,axy^4 + 5a^2x^3y$$

Case 3: When both dividend and divisor are polynomials.

Example: $\dfrac{x^4 + x^2y^2 + y^4}{x^2 + xy + y^2} = x^2 - xy + y^2$

Useful algebraic equations.

1. $a^2 - b^2 = (a + b)(a - b)$
2. $a^4 - b^4 = (a^2 + b^2)(a^2 - b^2) = (a^2 + b^2)(a + b)(a - b)$
3. $a^3 - b^3 = (a^2 + ab + b^2)(a - b)$
4. $a^3 + b^3 = (a^2 - ab + b^2)(a + b)$

5. $a^6 - b^6 = (a^3 + b^3)(a^3 - b^3) = (a^3 + b^3)$
$(a^2 + ab + b^2)(a - b)$
$= (a^3 + b^3)(a^3 - b^3) = (a^3 - b^3)$
$(a^2 - ab + b^2)(a + b)$
$= (a^3 + b^3)(a^3 - b^3) = (a^2 - b^2)$
$(a^4 + a^2b^2 + b^4)$
$= (a + b)(a - b)(a^2 + ab + b^2)(a^2 - ab + b^2)$

6. $\dfrac{a^2 - b^2}{a - b} = a + b$

7. $\dfrac{a^3 - b^3}{a - b} = a^2 + ab + b^2$

8. $\dfrac{a^3 + b^3}{a + b} = a^2 - ab + b^2$

9. $\dfrac{a^4 - b^4}{a + b} = a^3 - a^2b + ab^2 - b^3$

10. $\dfrac{a^5 - b^5}{a - b} = a^4 + a^3b + a^2b^2 + ab^3 + b^4$

11. $\dfrac{a^5 + b^5}{a + b} = a^4 - a^3b + a^2b^2 - ab^3 + b^4$

12. $\dfrac{a^6 - b^6}{a^2 - b^2} = a^4 + a^2b^2 + b^4$

Prime factors of algebraic polynomials.
1. $ab + ac + ad = a(b + c + d)$
2. $rs + rt + su + tu = (r + u)(s + t)$
3. $a^2 + 2ab + b^2 = (a + b)(a + b) = (a + b)^2$
4. $a^2 - 2ab + b^2 = (a - b)(a - b) = (a - b)^2$
5. $a^2 - b^2 = (a + b)(a - b)$
6. $ab - ac - ad = a(b - c - d)$
7. $a^3 - b^3 = (a - b)(a^2 + ab + b^2)$
8. $a^3 + b^3 = (a + b)(a^2 - ab + b^2)$
9. $a^2 + 3a + 2 = (a + 2)(a + 1)$
10. $a^2 - 3a + 2 = (a - 2)(a - 1)$
11. $a^2 + a - 2 = (a + 2)(a - 1)$
12. $a^2 - a - 2 = (a - 2)(a + 1)$
13. $a^4 + a^2b^2 + b^4 = (a^2 + ab + b^2)(a^2 - ab + b^2)$

List of algebraic fractions.

1. $\dfrac{a}{x} + \dfrac{b}{x} = \dfrac{a + b}{x}$

2. $\dfrac{b}{d} + \dfrac{c}{d} = \dfrac{b + c}{d}$

3. $\dfrac{1}{m} + \dfrac{4}{m} + \dfrac{7}{m} = \dfrac{12}{m}$

4. $\dfrac{b}{d} = \dfrac{ab}{ad}$

5. $\dfrac{a}{b} + \dfrac{c}{d} = \dfrac{ad + bc}{bd}$

6. $\dfrac{a}{cx} + \dfrac{b}{cy} = ay + \dfrac{bx}{cxy}$

7. $\dfrac{b}{d} - \dfrac{c}{d} = \dfrac{b - c}{d}$

8. $\dfrac{a}{b} - \dfrac{c}{d} = \dfrac{ad - bc}{bd}$

9. $\dfrac{a}{cx} - \dfrac{b}{cy} = \dfrac{ay - bx}{cxy}$

10. $\left(\dfrac{a}{b}\right)\left(\dfrac{c}{d}\right) = \dfrac{ac}{bd}$

11. $\left(\dfrac{a}{b} + \dfrac{c}{d}\right)\dfrac{e}{f} = \dfrac{ade + bce}{bdf}$

12. $\left(a + \dfrac{b}{a} - b\right)\left(a - \dfrac{b}{a^2} - b^2\right) = \dfrac{1}{a} - b$

13. $(b)\left(\dfrac{1}{b}\right) = \dfrac{b}{b} = 1$

14. $\dfrac{\frac{a}{b}}{\frac{x}{y}} = \left(\dfrac{a}{b}\right)\left(\dfrac{y}{x}\right) = \dfrac{ay}{bx}$

15. $\dfrac{\frac{a}{b}}{\frac{x}{y}} = \dfrac{\left(\frac{a}{1}\right)\left(\frac{x}{b}\right)}{y} = \left(\dfrac{a}{1}\right)\left(\dfrac{x}{b}\right)\left(\dfrac{1}{y}\right) = \dfrac{ax}{by}$

16. $\dfrac{\frac{a}{b}}{\frac{x}{y}} = \dfrac{\left(\frac{a}{1}\right)}{\left(\frac{b}{x}\right)\left(\frac{1}{y}\right)} = \left(\dfrac{a}{1}\right)\left(\dfrac{x}{b}\right)\left(\dfrac{y}{1}\right) = \dfrac{axy}{b}$

17. $\dfrac{\frac{ax}{1}}{\frac{b}{y}} = \left(\dfrac{ax}{1}\right)\left(\dfrac{y}{b}\right) = \dfrac{axy}{b}$

Reduction of fractions.

To reduce a fraction to its lowest terms, divide both the numerator and denominator by their greatest common factor and the result will be the fraction in its lowest terms.

Example: $\dfrac{a^2bc}{5a^2b^2} = \dfrac{c}{5b}$

To reduce a mixed quantity to an improper fraction, multiply the integral part of the denominator of the fraction and to the product add the numerator with its proper sign. The result placed over the denominator will give the improper fraction required.

Example: $\dfrac{a}{b} + 1 = \dfrac{a + b}{b}$

To reduce fractions to others equivalent and having a common denominator, multiply each of the numerators separately into all the denominators—except its own—to find the new numerators and all the denominators together for a common denominator.

Example: Reducing $\dfrac{a}{b}$ and $\dfrac{c}{d}$ to equivalent fractions having a common denominator yields $\dfrac{ad}{bd}$ and $\dfrac{bc}{bd}$.

Addition of fractions.

To reduce fractions to a common denominator, add the numerators together and subscribe the common denominator.

Example: $\dfrac{a}{b} + \dfrac{c}{d} = \dfrac{ad + cb}{bd}$

Subtraction of fractions.

To reduce fractions to a common denominator, subtract the numerator or the sum of the numerators of the fractions to be subtracted, from the numerator or the sum of the numerators of the others, and subscribe the common denominator.

Example: $\dfrac{a}{b} - \dfrac{c}{d} = \dfrac{ad - bc}{bd}$

Multiplication of fractions.

Multiply all the numerators together for a new numerator and all the denominators together for a new denominator.

Example: $\left(\dfrac{a}{b}\right)\left(\dfrac{c}{d}\right) = \dfrac{ac}{bd}$

Extracting the root of any degree of a monomial.

Extract the root of the numerical coefficient according to the rules of arithmetic and then divide the exponent of each letter by the index of the required root.

Example: $\sqrt[3]{64a^9b^3c^6} = 4a^3bc^2$

Addition and subtraction of radicals.

Add or subtract their coefficients and place the sum or difference as a coefficient before the common radical.

Example: $3\sqrt[3]{b} + 2\sqrt[3]{b} = 5\sqrt[3]{b}$

Multiplication and division of radicals.

Multiply or divide the quantities under the sign by each other and affect the result with the common radical.

Example: $\left(+\,2a\sqrt[3]{\dfrac{(a^2 + b^2)}{c}}\right)\left(-\,3a\sqrt[3]{\dfrac{(a^2 + b^2)^2}{d}}\right) =$

$-6a^2\sqrt[3]{\dfrac{(a^2 + b^2)^3}{cd}} = -\dfrac{6a^2\,(a^2 + b^2)}{\sqrt[3]{cd}}$

Rules of exponents in multiplication.

In order to multiply quantities expressed by the same letter, add the exponents.

Example: $(a^{3/4}b^{3/2}c^{-1})(a^2b^{2/3}c^{4/5}) = a^{11/4}b^{5/6}c^{1/5}$

Rules of exponents in division.

Subtract the exponent of the divisor from the exponent of the dividend.

Example: $\dfrac{a^{2/3}}{a^{-3/4}} = a^{2/3 - (-3/4)} = a^{17/12}$

Rules for raising a monomial to any power.

Multiply the exponent of the monomial by the exponent of the power required.

Example: $(a^{3/4})^5 = a^{3/4\,(5)} = a^{15/4}$

Binomial theorem.

The following section gives the binomial expressions on the left of the equal signs and their expansions on the right of the equal signs. Note that when a binomial quantity of the form $x + a$ is raised to any power, the successive terms are found in all cases to bear a certain relation to each other.

$$(x + a) = x + a$$
$$(x + a)^2 = x^2 + 2xa + x^2$$
$$(x + a)^3 = x^3 + 3x^2a + 3xa^2 + a^3$$
$$(x + a)^4 = x^4 + 4x^3a + 6x^2a^2 + 4xa^3 + a^4$$
$$(x + a)^5 = x^5 + 5x^4a + 10x^3a^2 + 10x^2a^3 + 5xa^4 + a^5$$
$$(x + a)^6 = x^6 + 6x^5a + 15x^4a^2 + 20x^3a^3 + 15x^2a^4$$
$$+ 6xa^5 + a^6$$
$$(x + a)^7 = x^7 + 7x^6a + 21x^5a^2 + 35x^4a^3 + 35x^3a^4$$
$$+ 21x^2a^5 + 7xa^6 + a^7$$
$$(x + a)^8 = x^8 + 8x^7a + 28x^6a^2 + 56x^5a^3 + 70x^4a^4$$
$$+ 56x^3a^5 + 28x^2a^6 + 8xa^7 + a^8$$

Inequalities.

Rule 1: an inequality is not changed in sense if the same number is added to, or subtracted from, both numbers.

Example: $8 > 5$, add 2 to the inequality.

$$\begin{array}{r} 8 > 5 \\ 2 = 2 \\ \hline 8 + 2 > 5 + 2 = 10 > 7 \end{array}$$

Rule 2: an inequality is not changed in sense if both members of the inequality are multiplied or divided by the same positive number.

Example: $7 > 4$, multiply the inequality by 2.

$$\begin{array}{r} 7 > 4 \\ 2 = 2 \\ \hline 14 > 8 \end{array}$$

Rule 3: an inequality is changed in sense if both members are multiplied or divided by the same negative number.

Example: $6 > 3$, multiply the inequality by -3.

$$\begin{array}{r} 6 > 3 \\ -3 = -3 \\ \hline -18 < -9 \end{array}$$

Rule 4: an inequality whose members are both positive is not changed in sense if both members are raised to the same positive power or if the same positive root of each member is taken.

Example: $5 > 4$, cube the inequality.

$$\begin{array}{r} 5 > 4 \\ 5^3 = 4^3 \\ \hline 125 > 64 \end{array}$$

Basic laws of analytical geometry

Introduction.
The following laws and equations include the ones frequently used in the solution of analytical geometry problems.

1. AXES
   The two fixed lines AX and AY are called axes.

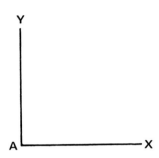

2. ABSCISSA
   The distance AM or NP of the point P from the axis AY is called the abscissa of the point P and is usually designated geometrically by the letter x.

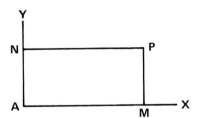

3. ORDINATE
   The distance AN or MP of the point P from the axis AX is called the ordinate of the point P and is usually designated geometrically by the letter y.

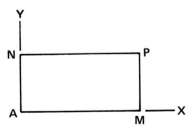

4. COORDINATES
   The two distances x and y are both designated the coordinates of the point P.

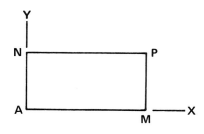

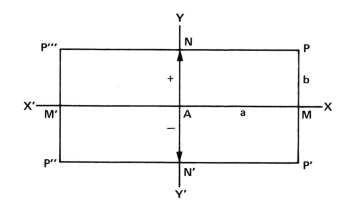

5. AXIS OF ABSCISSAS
The axis AX along which the abscissas are reckoned, is called the axis of abscissa or the axis of x's.

6. AXIS OF ORDINATES
The axis AY along which the ordinates are reckoned, is called the axis of ordinates or the axis of y's.

7. EQUATIONS OF A POINT
The characteristics of every point situated on the axis of y's is $x = 0$ and the characteristic of every point situated on the axis of x's is $y = 0$.

8. POSITIVE DISTANCES TO THE RIGHT
Positive distances such as AM are reckoned along AX to the right of point A.

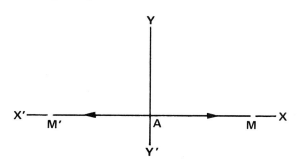

9. NEGATIVE DISTANCES TO THE LEFT
Negative distances such as AM′ are reckoned to the left of point A along AX′.

10. POSITIVE DISTANCES UPWARDS
Positive distances such as AN are reckoned along AY upwards from point A.
For P: $x = a$, $y = b$
For P′: $x = a$, $y = -b$
For P″: $x = -a$, $y = -b$
For P‴: $x = -a$, $y = b$

11. NEGATIVE DISTANCES DOWNWARDS
Negative distances such as AN′ are reckoned along AY downwards from point A.

12. SIGNS
Signs are either plus (+) or minus (−) and are used for expressions such as a and b, according to the position of the point in the plane of the axes AX and AY.

13. ABSOLUTE VALUES
The absolute or numerical values of the distances of the point from the two axes AX and AY.

14. ORIGIN OF COORDINATES
The point A is called the origin of coordinates since it is from this point that the distances are measured or reckoned.

15.
To determine the analytical expression for the distance between two given points that are situated in the same plane, let the coordinates of the first point P_1 be x', y' and the second point P_2 be x'', y''. Then:

$$P_1: \begin{array}{l} x = x' \\ y = y' \end{array}$$

$$P_2: \begin{array}{l} x = x'' \\ y = y'' \end{array}$$

Let the distance P_1P_2 be called R. Then:

$$R = \sqrt{(x' - x'')^2 + (y' - y'')^2}$$

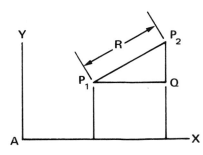

16. To find the equation of a straight line, the following equations of a straight line such as LS are given according to the different positions that it may assume.

$y = ax + b$	$x = c$
$y = ax - b$	$y = -ax + y$
$y = ax$	$y = ax - b$
$y = b$	$y = -ax - b$

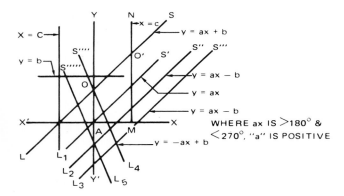

WHERE ax IS $>180°$ & $<270°$, "a" IS POSITIVE

17. To find the equation of a straight line that passes through a given point, the following equations are used:

$$y = ax + b \tag{1}$$
$$y' = ax' + b \tag{2}$$

Subtracting Eq. 2 from Eq. 1:

$$y - y' = a(x - x') \tag{3}$$

18. To find the equation of a straight line that passes through two given points, the following equations are used:

$$y = ax + b \tag{1}$$
$$y' = ax' + b \tag{2}$$
$$y'' = ax'' + b \tag{3}$$

Subtracting Eq. 3 from Eq. 2:

$$y' - y'' = a(x' - x'')$$

$$a = \frac{y' - y''}{x' - x''} \tag{4}$$

Again, subtracting Eq. 2 from Eq. 1:

$$y - y' = a(x - x')$$

Substituting in this equation the value of a obtained from Eq. 4:

$$y - y' = \frac{y' - y''}{x' - x''}(x - x') \tag{5}$$

19. To find the equation of a straight line parallel to a given straight line, let the equation of the given straight line SL be:

$$y = ax + b$$

where a is the tangent of the angle OTA. Then:

$$y = ax + b'$$

where $b' = AO'$

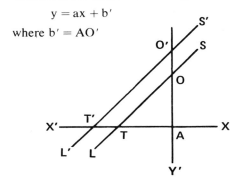

20. To find the equation of a straight line drawn through a given point perpendicular to a given straight line use:

$$y - y' = -\frac{1}{a}(x - x')$$

21. To find the equation to a circle, the following equations are given where A is a circle whose center is at O and whose radius is OP. Also, the coordinates of the point O are x'y' and of any point P in the circumference x, y.

$$\text{OP}^2 \text{ or } \text{R}^2 = (x - x')^2 + (y - y')^2$$
$$y^2 = 2rx - x^2$$
$$x^2 = 2ry - y^2$$
$$r^2 = x^2 + y^2$$
$$\text{R}^2 = \text{Radius of the circle}$$

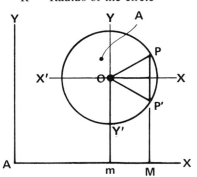

Fillet area (right-angle fillet)

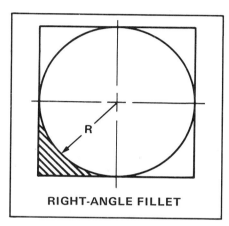

RIGHT-ANGLE FILLET

Introduction.

Frequently, design solutions require the calculation of the cross-sectional area of a right-angle fillet. This cross-sectional area is given by the following equations:

$$A = R^2 - \frac{\pi R^2}{4} \qquad (1)$$

$$A = 0.215 R^2 \qquad (2)$$

Nomenclature:

A = Cross-sectional area of a right-angle fillet, sq in
R = Radius, inches

To facilitate the solution of Eqs. 1 and 2, Table 1 can be used. Values for R range from $\frac{1}{32}$ to 5 inches in Table 1.

Example 1:

Determine the cross-sectional area of a right-angle fillet having a radius of 2.0 inches.

Solution 1:

Locate 2.0 inches in the radius column of Table 1 and read the answer of A = 0.858 sq in.

RADIUS, R, (inches)	AREA, A, (sq in)
1/32	0.00021
1/16	0.00083
3/32	0.00188
1/8	0.00335
5/32	0.00523
3/16	0.00754
7/32	0.01027
1/4	0.01341
5/16	0.02095
3/8	0.03018
7/16	0.04107
1/2	0.0536
9/16	0.0679
5/8	0.0838
11/16	0.1014
3/4	0.1207
13/16	0.1414
7/8	0.1643
15/16	0.1886
1	0.2146
1-1/6	0.2422
1-1/8	0.2641
1-3/16	0.3026
1-1/4	0.3378
1-5/16	0.3697
1-3/8	0.4057
1-7/16	0.4434
1-1/2	0.4828
1-9/16	0.5239
1-5/8	0.5667
1-11/16	0.6111
1-3/4	0.6572
1-13/16	0.7049
1-7/8	0.7543
1-15/16	0.8056
2	0.858
2-1/8	0.969
2-1/4	1.086
2-3/8	1.210
2-1/2	1.341
2-5/8	1.478
2-3/4	1.623
2-7/8	1.774
3	1.931
3-1/4	2.267
3-1/2	2.629
3-3/4	3.018
4	3.434
4-1/4	3.876
4-1/2	4.346
4-3/4	4.842
5	5.365
TABLE I	

Fillet area
(general fillet)

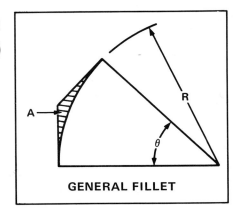

GENERAL FILLET

Introduction.

Frequently, design solutions require the calculation of the cross-sectional area of a general fillet. This cross-sectional area is given by the following equations:

$$A = R^2 \left(\tan \frac{\theta}{2} - \frac{\pi \theta}{360} \right) \tag{1}$$

$$A = R^2 f(\theta) \tag{2}$$

Nomenclature:

A = Cross-sectional area of a general fillet, sq in

R = Radius, inches

θ = Angle, deg

$f(\theta)$ = Area function of θ, dimensionless

Example 1:

Determine the cross-sectional area of a fillet having a radius of 2.0 inches and angle of 23 deg.

Solution 1:

Locate 23 deg in the θ(deg) column of Table 1 and read 0.00274 from the $f(\theta)$ column. Substituting this value into Eq. 2 we arrive at:

$A = R^2 f(\theta)$

$A = (2.0^2) (0.00274)$

$A = 0.01096$ sq in

θ (deg)	f (θ)	θ (deg)	f (θ)	θ (deg)	f (θ)	$\dot{\theta}$ (deg)	f (θ)
		46	0.02305	91	0.22348	136	1.28826
		47	0.02466	92	0.23268	137	1.34310
		48	0.02635	93	0.24220	138	1.40081
		49	0.02812	94	0.25206	139	1.46162
5	0.00003	50	0.02998	95	0.26228	140	1.52575
6	0.00005	51	0.03192	96	0.27285	141	1.59346
7	0.00008	52	0.03395	97	0.28381	142	1.66503
8	0.00011	53	0.03607	98	0.29516	143	1.74077
9	0.00016	54	0.03829	99	0.30691	144	1.82105
10	0.00022	55	0.04060	100	0.31909	145	1.90623
11	0.00030	56	0.04302	101	0.33171	146	1.99676
12	0.00038	57	0.04554	102	0.34478	147	2.09313
13	0.00049	58	0.04816	103	0.35833	148	2.19587
14	0.00061	59	0.05090	104	0.37237	149	2.30561
15	0.00075	60	0.05375	105	0.38693	150	2.42305
16	0.00091	61	0.05672	106	0.40202	151	2.54899
17	0.00110	62	0.05981	107	0.41767	152	2.68433
18	0.00130	63	0.06302	108	0.43390	153	2.83012
19	0.00154	64	0.06636	109	0.45074	154	2.98757
20	0.00179	65	0.06984	110	0.45822	155	3.15808
21	0.00208	66	0.07345	111	0.48635	156	3.34327
22	0.00239	67	0.07720	112	0.50518	157	3.54507
23	0.00274	68	0.08110	113	0.52472	158	3.76574
24	0.00312	69	0.08514	114	0.54503	159	4.00798
25	0.00353	70	0.08934	115	0.56612	160	4.27502
26	0.00398	71	0.09370	116	0.58804	161	4.57077
27	0.00446	72	0.09822	117	0.61083	162	4.90003
28	0.00498	73	0.10292	118	0.63454	163	5.26871
29	0.00554	74	0.10778	119	0.65919	164	5.68420
30	0.00615	75	0.11283	120	0.68485	165	6.15586
31	0.00680	76	0.11806	121	0.71157	166	6.69572
32	0.00749	77	0.12348	122	0.73940	167	7.31954
33	0.00823	78	0.12911	123	0.76839	168	8.04829
34	0.00902	79	0.13493	124	0.79862	169	8.91059
35	0.00987	80	0.14097	125	0.83015	170	9.94652
36	0.01076	81	0.14722	126	0.86305	171	11.21395
37	0.01171	82	0.15370	127	0.89741	172	12.79968
38	0.01272	83	0.16041	128	0.93329	173	14.84015
39	0.01378	84	0.16737	129	0.97081	174	17.56270
40	0.01490	85	0.17457	130	1.01004	175	21.37660
41	0.01609	86	0.18202	131	1.05111	176	27.10036
42	0.01734	87	0.18975	132	1.09412	177	36.64384
43	0.01866	88	0.19774	133	1.13920	178	55.73662
44	0.02005	89	0.20603	134	1.18648	179	113.02658
45	0.02151	90	0.21460	135	1.23612	180	∞

TABLE 1

Areas of a few common shapes

Introduction.
This article is a quick reference to help you find the areas of a few common cross-sections. The equations have been simplified to avoid terms containing pi and the dimensions are measured easily across the shape.

Example 1:
Given a square that has a side of 4.0 inches, determine its area.

Solution 1:
Solving the area equation for a square:
Area = a^2
Area = 4.0^2
Area = 16.0 sq in

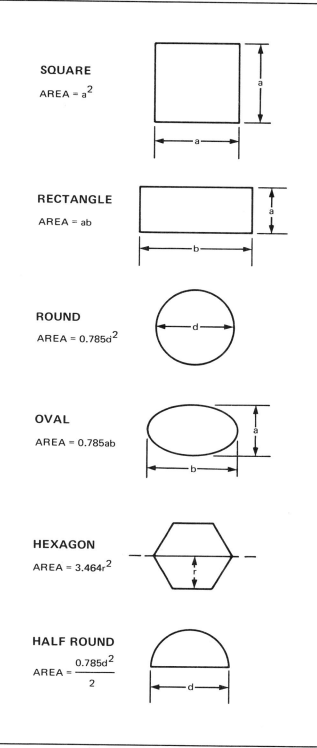

SQUARE

AREA = a^2

RECTANGLE

AREA = ab

ROUND

AREA = $0.785d^2$

OVAL

AREA = 0.785ab

HEXAGON

AREA = $3.464r^2$

HALF ROUND

AREA = $\dfrac{0.785d^2}{2}$

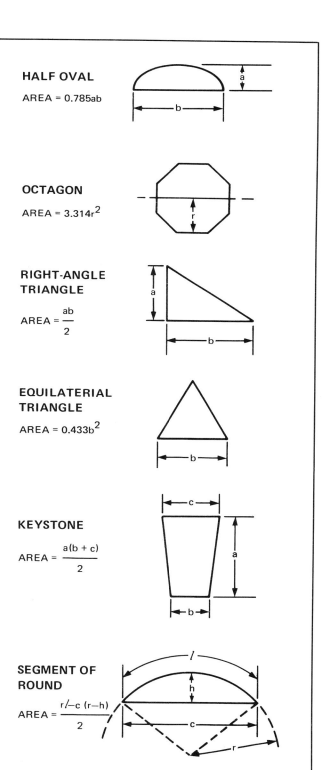

HALF OVAL

AREA = 0.785ab

OCTAGON

AREA = $3.314r^2$

RIGHT-ANGLE TRIANGLE

AREA = $\dfrac{ab}{2}$

EQUILATERIAL TRIANGLE

AREA = $0.433b^2$

KEYSTONE

AREA = $\dfrac{a(b + c)}{2}$

SEGMENT OF ROUND

AREA = $\dfrac{rl-c\,(r-h)}{2}$

Area of circle

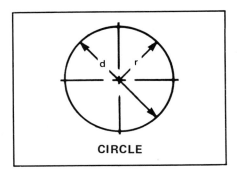

CIRCLE

Introduction.
To determine the area, radius, diameter and circumference of a circle the following equations can be used:

Area: $A = \pi r^2$
$A = 3.1416r^2$
$A = 0.7854d^2$

Radius: $r = \dfrac{C}{6.2832}$

Diameter: $d = \dfrac{C}{3.1416}$

Circumference: $C = 2\pi r$
$C = 6.2832r$
$C = 3.1416d$

Nomenclature:
A = Area, sq in
r = Radius, inches
d = Diameter, inches
C = Circumference, inches

Nomogram.
A nomogram can be used to expedite the solution of the above equations. When using the nomogram, the following procedure should be followed:

To find the area of a circle use scales 1, 2 and 5.
To find the radius of a circle use scales 1, 4 and 7.
To find the diameter of a circle use scales 1, 4 and 7.
To find the circumference of a circle use scales 1, 3 and 6.

Example 1:
Given a circle that has a radius of 1.4 inches, determine its area.
Solution 1:
Construct a line from 1.4 on scale 5 to $A = 3.1416r^2$ on scale 1 and where this line intersects scale 2, read the answer of 6.2 sq in.

Example 2:
Given a circle that has a circumference of 3.50 inches, determine its radius.
Solution 2:
Construct a line from 3.50 on scale 7 to $r = C/6.2832$ on scale 1 and where this line intersects scale 4 read the answer of 0.56 inches.

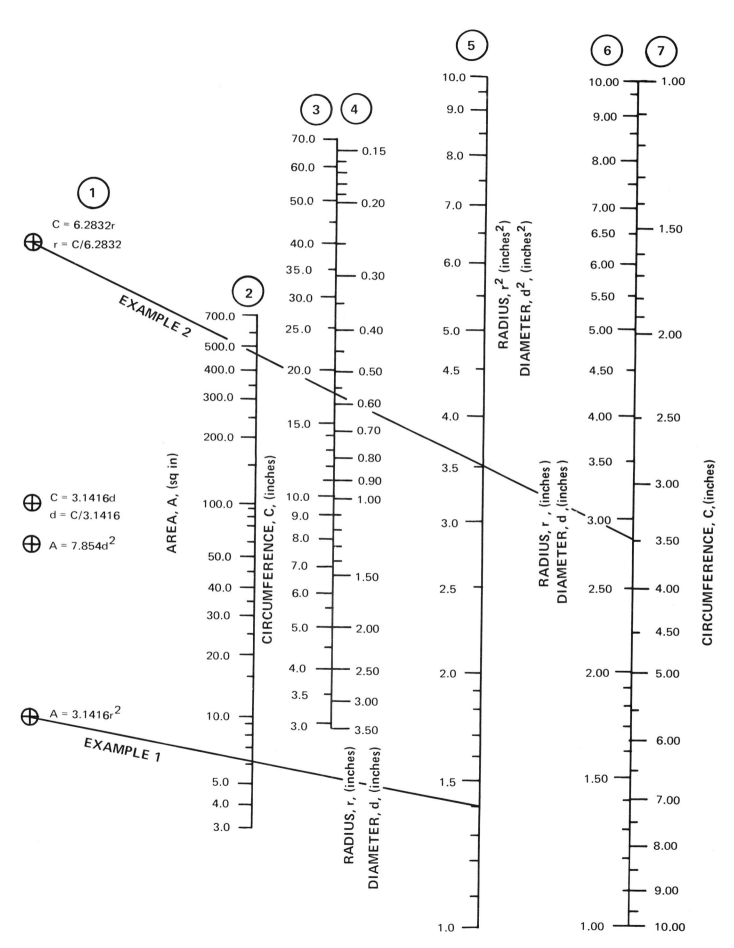

① C = 6.2832r
   r = C/6.2832

EXAMPLE 2

② C = 3.1416d
   d = C/3.1416

③ A = 7.854d²

④ A = 3.1416r²

EXAMPLE 1

③ ④

⑤

⑥ ⑦

AREA, A, (sq in)

CIRCUMFERENCE, C, (inches)

RADIUS, r, (inches)
DIAMETER, d, (inches)

RADIUS, r² (inches²)
DIAMETER, d², (inches²)

RADIUS, r, (inches)
DIAMETER, d, (inches)

CIRCUMFERENCE, C, (inches)

Area of circular sector

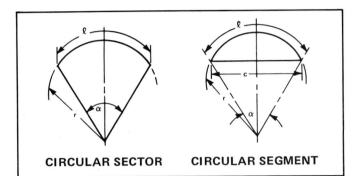

CIRCULAR SECTOR   CIRCULAR SEGMENT

Introduction.

To determine the area, radius, length of arc and angle in a circular sector and the radius, length of arc and angle in a circular segment, the following equations can be used:

$$\text{Circular sector: } A = \tfrac{1}{2}r\ell$$
$$A = 0.008727 \; \alpha \, r^2$$
$$r = \frac{2A}{\ell}$$
$$r = \frac{57.296\ell}{\alpha}$$
$$\ell = \frac{2A}{r}$$
$$\ell = 0.01745r \; \alpha$$
$$\alpha = \frac{57.296\ell}{r}$$

$$\text{Circular segment: } r = \frac{2A}{\ell}$$
$$r = \frac{57.296\ell}{\alpha}$$
$$\ell = \frac{2A}{r}$$
$$\ell = 0.01745r \; \alpha$$
$$\alpha = \frac{57.296\ell}{r}$$

Nomenclature:
- A = Area, sq in
- r = Radius, inches
- ℓ = Length of arc, inches
- α = Angle, deg

Nomogram.

A nomogram can be used to expedite the solution of the above equations. When using the nomogram, the following procedure should be followed:

To find the area and length ($\ell = 2A/r$) of a circular sector and the length ($\ell = 2A/r$) of a circular segment, use scales 1, 3 and 4.

To find the length ($\ell = 0.01745r \; \alpha$) of a circular sector and circular segment, use scales 1, 2 and 5.

To find the radius and angle of a circular sector and circular segment, use scales 1, 2 and 5.

Example 1:

Given a circular sector that has a radius of 2.50 inches and length of 1.50 inches, determine its area.

Solution 1:

Construct a line from 2.50 on scale 1 to 1.50 on scale 4 and where this line intersects scale 3 read the answer of 1.87 sq in.

Example 2:

Given a circular sector that has an area of 20.0 sq in and a radius of 4.00 inches, determine its length.

Solution 2:

Construct a line from 4.00 on scale 1 to 20.0 on scale 3 and continue this line until it intersects scale 4. At this intersection read the answer of ℓ = 10.00 inches.

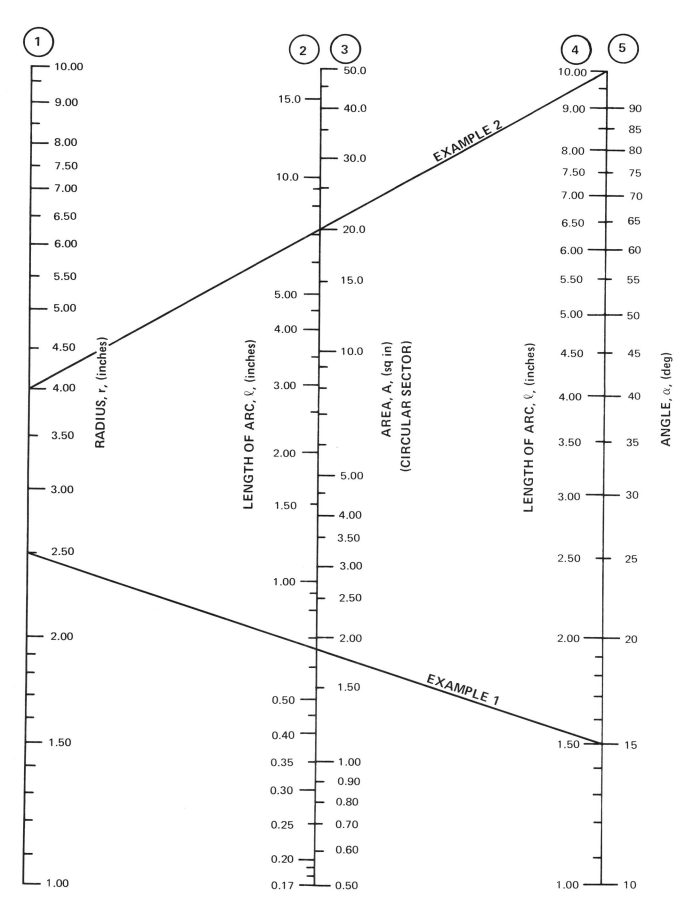

① RADIUS, r, (inches)

② LENGTH OF ARC, ℓ, (inches)

③ AREA, A, (sq in) (CIRCULAR SECTOR)

④ LENGTH OF ARC, ℓ, (inches)

⑤ ANGLE, α, (deg)

EXAMPLE 2

EXAMPLE 1

Area of cone

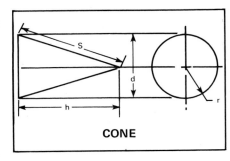

CONE

Introduction.
To determine the area, radius, diameter, side and height of a cone, the following equations can be used:

$$A = 3.1416rS$$
$$A = 1.5708dS$$
$$S = \sqrt{r^2 + h^2}$$

Nomenclature:
A = Area of conical surface, sq in
S = Length parallel to side of cone, inches
h = Height, inches
r = Radius, inches
d = Diameter, inches

Nomogram.
A nomogram can be used to expedite the solution of the above equations. When using the nomogram, the following procedure should be followed:
To find the area of conical surface use scales 2, reference line, 5, 6 and 7.
To find the length parallel to side of cone use scales 1, 3 and 4.

Example 1:
Given a cone that has a diameter of 4.00 inches and length parallel to side of cone as 10.0 inches, determine its area.
Solution 1:
Construct a line from A = 1.5708dS on scale 2 to 4.00 on scale 6 and where this line intersects the reference line, label point A. From point A construct a line to 10.0 on scale 7 and where this line intersects scale 5 read the answer of 63.0 sq in.
Example 2:
Given a cone that has a radius of 3.50 inches and height of 2.00 inches, determine the length parallel to side of cone.
Solution 2:
Construct a line from 3.50 on scale 1 to 2.00 on scale 4 and where this line intersects scale 3 read the answer of 4.00 inches.

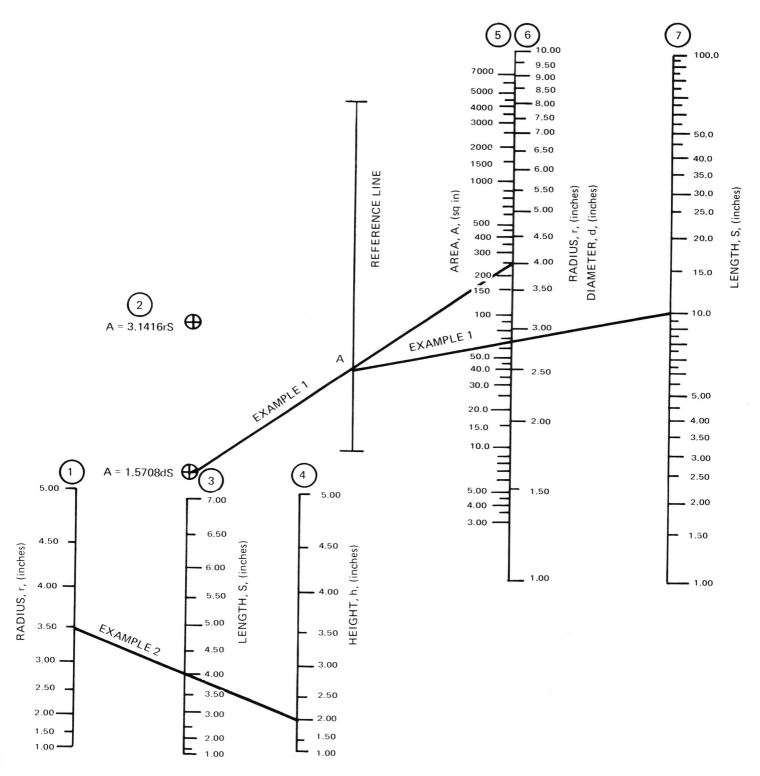

REFERENCE LINE

⑤ ⑥ ⑦

AREA, A, (sq in)

RADIUS, r, (inches)
DIAMETER, d, (inches)

LENGTH, S, (inches)

EXAMPLE 1

② A = 3.1416rS ⊕

A = 1.5708dS ⊕ ③ ④ ①

EXAMPLE 1

A

RADIUS, r, (inches)

LENGTH, S, (inches)

HEIGHT, h, (inches)

EXAMPLE 2

Area of cycloid

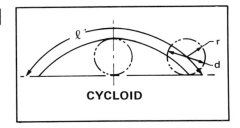

CYCLOID

Introduction:

To determine the area and length of a cycloid the following equations can be used:

$$A = 3\pi r^2$$
$$A = 9.4248r^2$$
$$A = 2.3562d^2$$
$$\ell = 8r$$
$$\ell = 4d$$

Nomenclature:

A = Area enclosed between the curve and base line, sq in

ℓ = Length of the cycloidal curve, inches

r = Radius of generating circle, inches

d = Diameter of generating circle, inches

Nomogram.

A nomogram can be used to expedite the solution of the above equations. When using the nomogram the following procedure should be followed:

To find the area enclosed between the curve and base line use scales 1, 4 and 5.

To find the length of the cycloidal curve use scales 2, 3 and 6.

Example 1:

Given that the radius of the generating circle is 1.10 inches, determine the area enclosed between the curve and base line.

Solution 1:

Construct a line from 1.10 on scale 5 to A = 9.4248r² on scale 4 and continue this line until it intersects scale 1. At this intersection point read the answer of 11.40 sq in.

Example 2:

Given that the diameter of the generating circle is 1.50 inches, determine the length of the cycloid curve.

Solution 2:

Construct a line from 1.50 on scale 6 to ℓ = 4.00d on scale 2 and where this line intersects scale 3 read the answer of 6.0 inches.

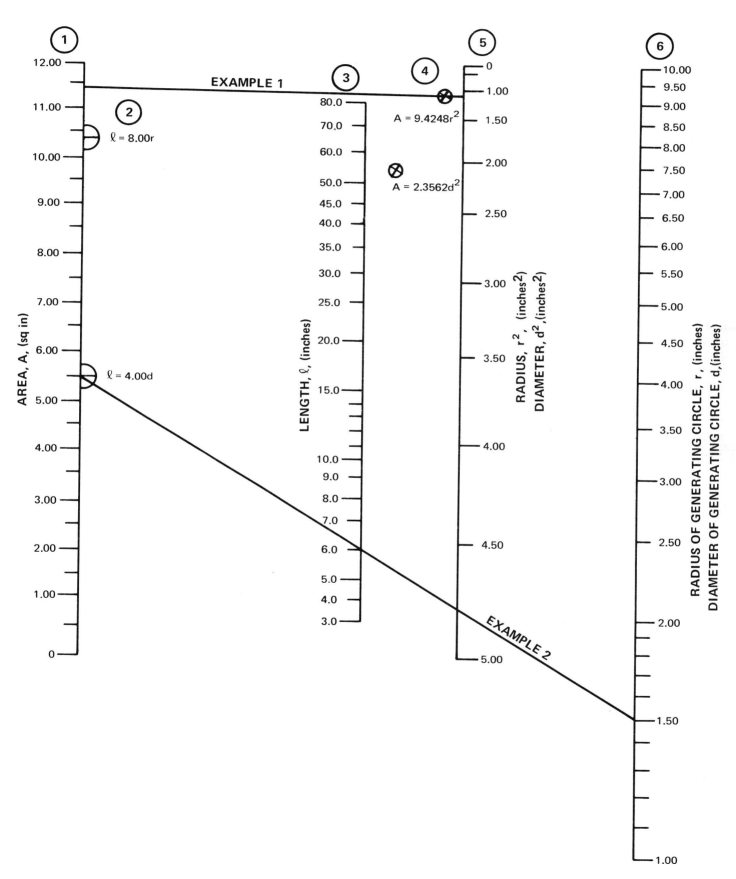

Area of cylindrical surface

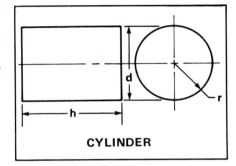

CYLINDER

Introduction.
To determine the area of a cylindrical surface the following equations can be used:

$$A = 6.2832rh$$
$$A = 3.1416dh$$

Nomenclature:

A = Area of cylindrical surface, sq in

r = Radius, inches

h = Height, inches

d = Diameter, inches

Nomogram.
A nomogram can be used to expedite the solution of the above equations.

Example 1:
Given a cylindrical surface that has a diameter of 5.00 inches and height of 10.0 inches, determine its area.

Solution 1:
Construct a line from A = 3.1416dh on scale 1 to 5.00 on scale 3 and where this line intersects the reference line, label point A. From point A construct a line to 10.0 on scale 4 and where this line intersects scale 2 read the answer of 157 sq in.

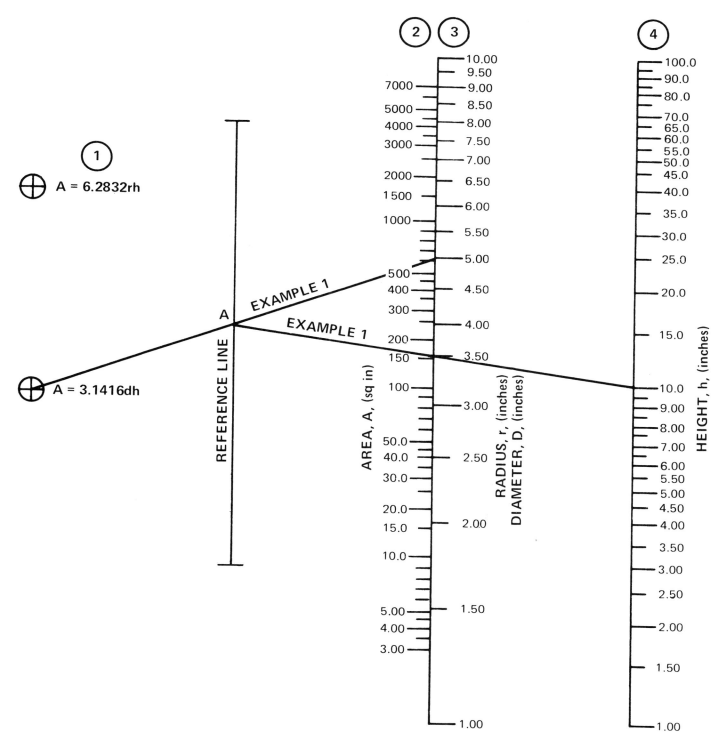

① A = 6.2832rh

① A = 3.1416dh

A

REFERENCE LINE

EXAMPLE 1

EXAMPLE 1

② ③ ④

AREA, A, (sq in)

RADIUS, r, (inches)
DIAMETER, D, (inches)

HEIGHT, h, (inches)

Area of ellipse

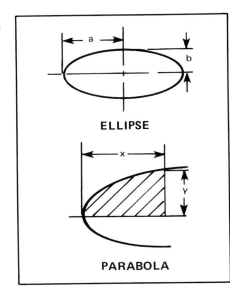

ELLIPSE

PARABOLA

Introduction.
To determine the area of an ellipse and parabola, the following equations can be used:

Ellipse: $A = \pi ab$
$A = 3.1416ab$

Parabola: $A = \frac{2}{3}xy$

Nomenclature:
A = Total area of ellipse, sq in; shaded area of parabola, sq in
a = Major or larger axis, inches
b = Minor or smaller axis, inches
x = Horizontal axis, inches
y = Vertical axis, inches

Nomogram.
A nomogram can be used to expedite the solution of the above equations.

Example 1:
Given an ellipse that has a major axis of 10.00 inches and minor axis of 5.00 inches, determine its area.

Solution 1:
Construct a line from 10.00 on scale 1 to 5.00 on scale 4 and where this line intersects scale 3 read the answer of 160.0 sq in.

Example 2:
Given a parabola that has a horizontal axis of 3.00 inches and vertical axis of 2.00 inches, determine its area.

Solution 2:
Construct a line from 3.00 on scale 1 to 2.00 on scale 4 and where this line intersects scale 2 read the answer of 4.00 sq in.

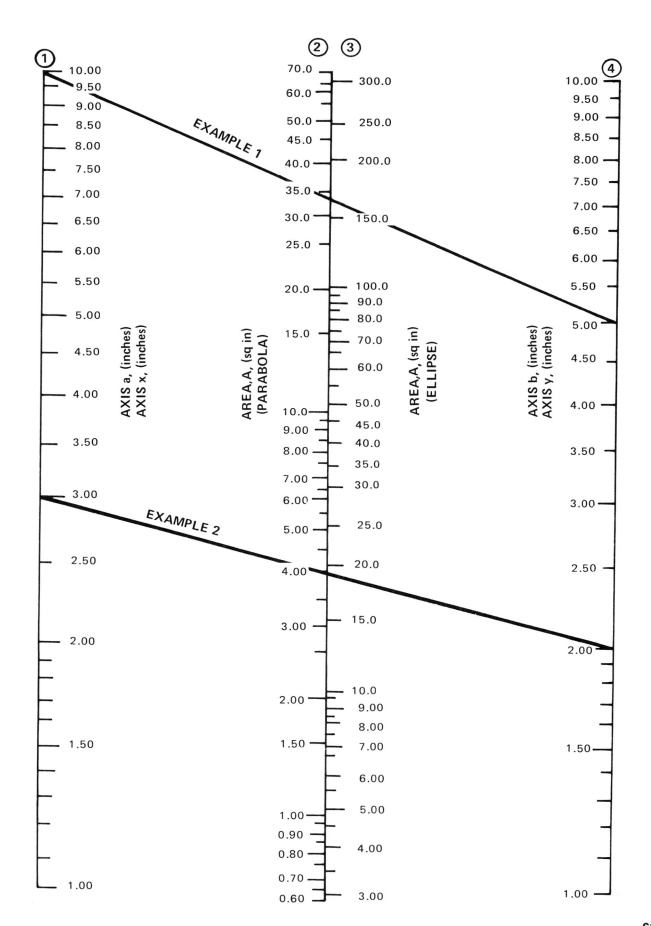

Area of rectangle

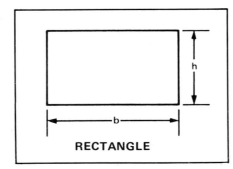

RECTANGLE

Introduction.
To determine the area of a rectangle the following equation can be used:
$$A = bh$$

Nomenclature:
A = Area, sq in
b = Base, inches
h = Height, inches

Nomogram.
A nomogram can be used to expedite the solution of the above equation.

Example 1:
Given a rectangle that has a base of 5.00 inches and height of 3.00 inches, determine its area.

Solution 1:
Construct a line from 5.00 on scale 1 to 3.00 on scale 3 and where this line intersects scale 2 read the answer of 15.0 sq in.

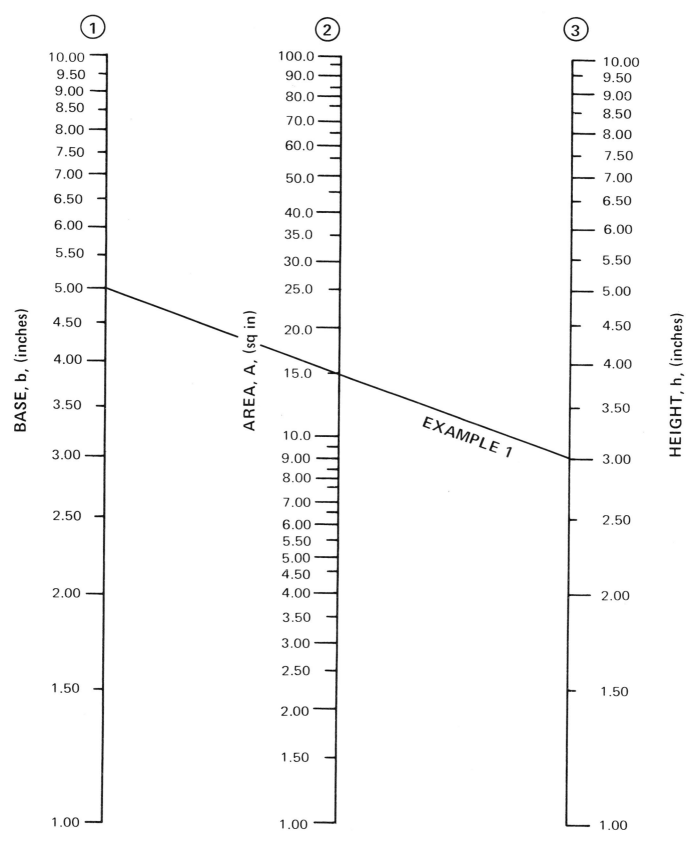

Area of regular hexagon

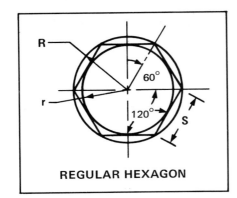

REGULAR HEXAGON

Introduction.

To determine the area, radius of circumscribed circle and radius of inscribed circle of a regular hexagon, the following equations can be used:

Area of regular hexagon:	$A = 2.598S^2$
	$A = 2.598R^2$
	$A = 3.464r^2$
Radius of circumscribed circle:	$R = 1.155r$
	$S = 1.155r$
Radius of inscribed circle:	$r = 0.866S$
	$r = 0.866R$
Side $S = R = 1.155r$	

Nomenclature:

A = Area, sq in
R = Radius of circumscribed circle, inches
r = Radius of inscribed circle, inches
S = Side S, inches

Nomogram.

A nomogram can be used to expedite the solution of the above equations. When using the nomogram, the following procedure should be followed:

To find the area of a regular hexagon, use scales 4, 5 and 6.

To find the radius of a circumscribed circle of a regular hexagon, use scales 1, 2 and 3.

To find the radius of an inscribed circle of a regular hexagon, use scales 1, 2 and 3.

Example 1:

Given radius of a circumscribed circle of a regular hexagon as being 1.50 inches, determine the area of the regular hexagon.

Solution 1:

Construct a line from 1.50 on scale 6 to A = 2.598R² on scale 5 and continue this line until it intersects scale 4 and at this intersection point read the answer of 5.85 sq in.

Example 2:

Given sides of a regular hexagon as 2.00 inches, determine the radius of the inscribed circle of a regular hexagon.

Solution 2:

Construct a line from 2.00 on scale 3 to r = 0.866S on scale 1 and where this line intersects scale 2 read the answer of 1.73 inches.

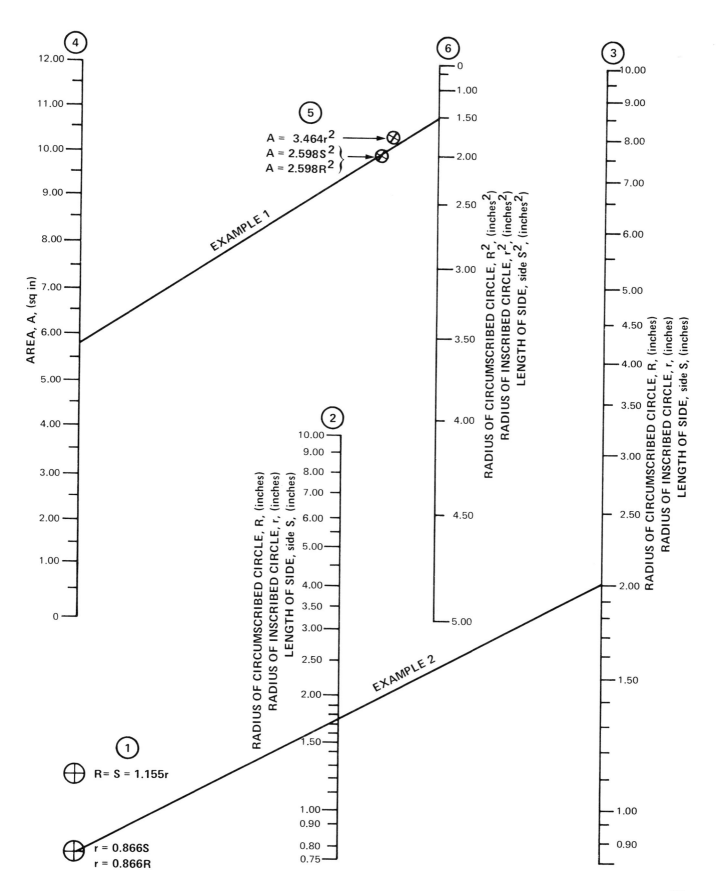

Area of regular octagon

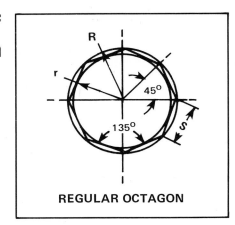

REGULAR OCTAGON

Introduction.

To determine the area, radius of circumscribed circle and radius of inscribed circle of a regular octagon, the following equations can be used:

Area of regular octagon: $A = 4.828S^2$
$A = 2.828R^2$
$A = 3.314r^2$
Radius of circumscribed $R = 1.307S$
circle: $R = 1.082r$
Radius of inscribed $r = 1.207S$
circle: $r = 0.924R$
Side $S = 0.765R = 0.828r$

Nomenclature:

A = Area, sq in
R = Radius of circumscribed circle, inches
r = Radius of inscribed circle, inches
S = Side S, inches

Nomogram.

A nomogram can be used to expedite the solution of the above equations.

Example 1:

Given radius of a circumscribed circle of a regular octagon as being 2.00 inches, determine the area of the regular octagon.

Solution 1:

Construct a line from 2.00 on scale 6 to $A = 2.828R^2$ on scale 4 and where this line intersects scale 5 read the answer of 11.30 sq in.

Example 2:

Given side S of a regular octagon as 1.60 inches, determine the radius of the circumscribed circle of a regular octagon.

Solution 2:

Construct a line from 1.60 on scale 3 to $R = 1.307S$ on scale 1 and where this line intersects scale 2 read the answer of 2.09 inches.

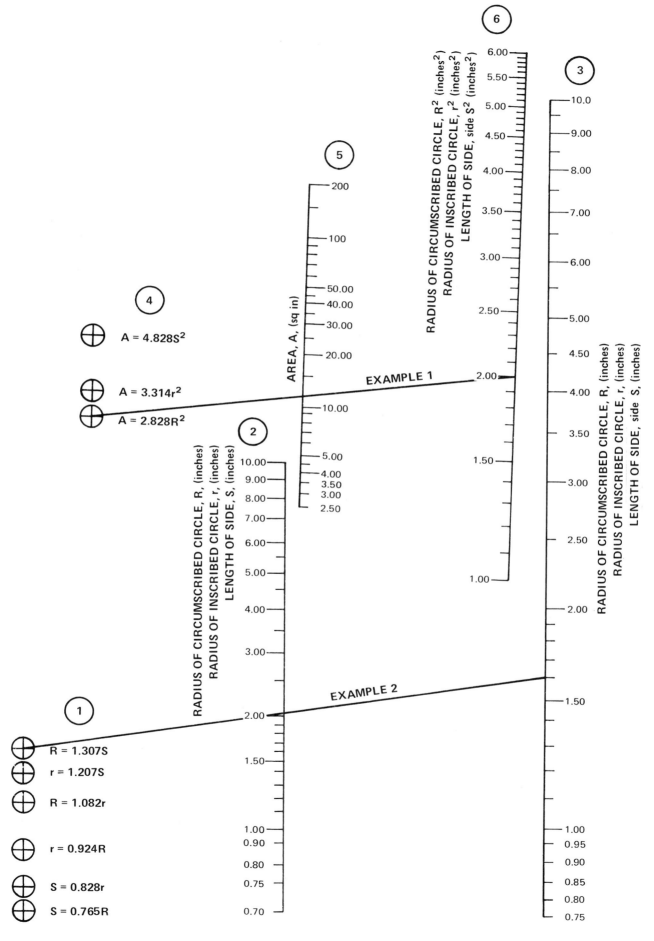

RADIUS OF CIRCUMSCRIBED CIRCLE, R^2 (inches²)
RADIUS OF INSCRIBED CIRCLE, r^2 (inches²)
LENGTH OF SIDE, side S^2 (inches²)

⑥

6.00
5.50
5.00
4.50
4.00
3.50
3.00

2.50

2.00

1.50

1.00

③

RADIUS OF CIRCUMSCRIBED CIRCLE, R, (inches)
RADIUS OF INSCRIBED CIRCLE, r, (inches)
LENGTH OF SIDE, side S, (inches)

10.0
9.00
8.00
7.00
6.00
5.00
4.50
4.00
3.50
3.00
2.50
2.00
1.50
1.00
0.95
0.90
0.85
0.80
0.75

⑤

AREA, A, (sq in)

200
100
50.00
40.00
30.00
20.00
10.00
5.00
4.00
3.50
3.00
2.50

EXAMPLE 1

④

⊕ A = 4.828S²
⊕ A = 3.314r²
⊕ A = 2.828R²

②

RADIUS OF CIRCUMSCRIBED CIRCLE, R, (inches)
RADIUS OF INSCRIBED CIRCLE, r, (inches)
LENGTH OF SIDE, S, (inches)

10.00
9.00
8.00
7.00
6.00
5.00
4.00
3.00
2.00

EXAMPLE 2

1.50
1.00
0.90
0.80
0.75
0.70

①

⊕ R = 1.307S
⊕ r = 1.207S
⊕ R = 1.082r
⊕ r = 0.924R
⊕ S = 0.828r
⊕ S = 0.765R

RADIUS OF CIRCUMSCRIBED CIRCLE, R, (inches)
RADIUS OF INSCRIBED CIRCLE, r, (inches)
LENGTH OF SIDE, side S, (inches)

2.00

1.50

1.00
0.95
0.90
0.85
0.80
0.75

Area of regular polygon

Introduction.

To determine the area of a regular polygon, the following equations can be used:

$$A = \frac{Snr}{2}$$

$$A = nr^2 \tan\left(\frac{180°}{n}\right)$$

Nomenclature:

A = Area, sq in
S = Length of one side, inches
n = Number of sides
r = Radius of inscribed circle, inches

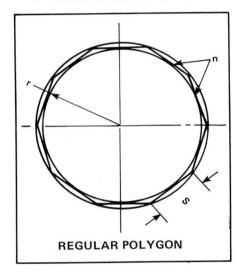

REGULAR POLYGON

Nomogram.

A nomogram can be used to expedite the solution of the above equations.

Example 1:

Given a regular polygon that has a radius of the inscribed circle of 20.0 inches, 10 sides and the length of each side is 12.9 inches, determine its area.

Solution 1:

Solving for $\tan\left(\frac{180°}{n}\right)$:

$$\left(\frac{180°}{10}\right) = 18°$$

The tan of 18° from Table 1 is equal to 0.324.

Construct a line from 20.0 on scale 1 to 10 on scale 3 and where this line intersects the reference line, label point A. Construct a line from point A to 0.324 on scale 4 and where this line intersects scale 2 read the answer of 1,300 sq in.

$\tan\left(\dfrac{180°}{n}\right)$			
Degree	**Decimal Value**	**Degree**	**Decimal Value**
0	0.000	46	1.036
1	0.017	47	1.072
2	0.034	48	1.111
3	0.052	49	1.150
4	0.069	50	1.192
5	0.087	51	1.235
6	0.105	52	1.280
7	0.122	53	1.327
8	0.140	54	1.376
9	0.158	55	1.428
10	0.176	56	1.483
11	0.194	57	1.540
12	0.212	58	1.600
13	0.230	59	1.664
14	0.249	60	1.732
15	0.267	61	1.804
16	0.286	62	1.881
17	0.305	63	1.963
18	0.324	64	2.050
19	0.344	65	2.145
20	0.364	66	2.246
21	0.383	67	2.356
22	0.404	68	2.475
23	0.424	69	2.605
24	0.445	70	2.747
25	0.466	71	2.904
26	0.487	72	3.078
27	0.509	73	3.271
28	0.531	74	3.487
29	0.554	75	3.732
30	0.577	76	4.011
31	0.600	77	4.331
32	0.624	78	4.705
33	0.649	79	5.145
34	0.674	80	5.671
35	0.700	81	6.314
36	0.726	82	7.115
37	0.753	83	8.144
38	0.781	84	9.514
39	0.809	85	11.430
40	0.839	86	14.300
41	0.869	87	19.080
42	0.900	88	28.640
43	0.932	89	57.290
44	0.965		
45	1.000		

TABLE 1

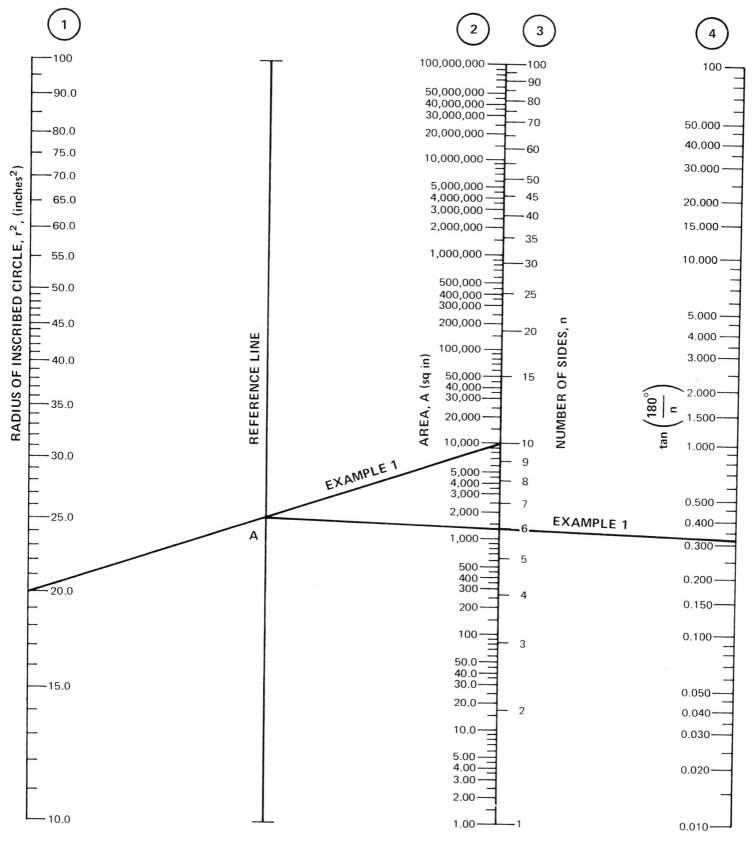

Area of triangle

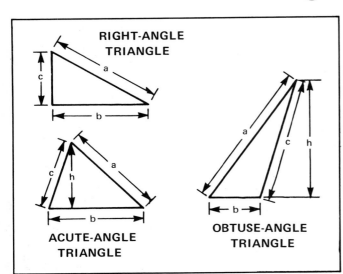

RIGHT-ANGLE TRIANGLE

ACUTE-ANGLE TRIANGLE

OBTUSE-ANGLE TRIANGLE

Introduction.
To determine the area of a right, acute and obtuse-angle triangle and the hypotenuse on a right-angle triangle, the following equations can be used:

Right-angle triangle: $A = \dfrac{1}{2}\, bc$

Acute-angle triangle: $A = \dfrac{1}{2}\, bh$

Obtuse-angle triangle: $A = \dfrac{1}{2}\, bh$

In a right-angle triangle,
side a or hypotenuse: $a = \sqrt{b^2 + c^2}$

Nomenclature:
A = Area, sq in
a = side a or hypotenuse, inches
b = side b, inches
c = side c, inches
h = height, inches

Nomogram:
A nomogram can be used to expedite the solution of the above equations. When using the nomogram the following procedure should be followed.

To find side a or hypotenuse of a right-angle triangle use scales 2, 3 and 5.

To find the area of a right-angle triangle use scales 1, 4 and 6.

Example 1:
Given a right-angle triangle that has a side b of 10.0 inches and side c of 4.00 inches, determine its area.

Solution 1:
On the nomogram construct a line from 4.00 on scale 6 to 10.0 on scale 1 and where this line intersects scale 4 read the answer of 20.0 sq in.

Example 2:
Given a right-angle triangle that has a side b of 4.00 inches and side c of 3.00 inches, determine the hypotenuse (side a).

Solution 2:
On the nomogram construct a line from 4.00 on scale 2 to 3.00 on scale 5 and where this line intersects scale 3 read the answer of 5.00 inches.

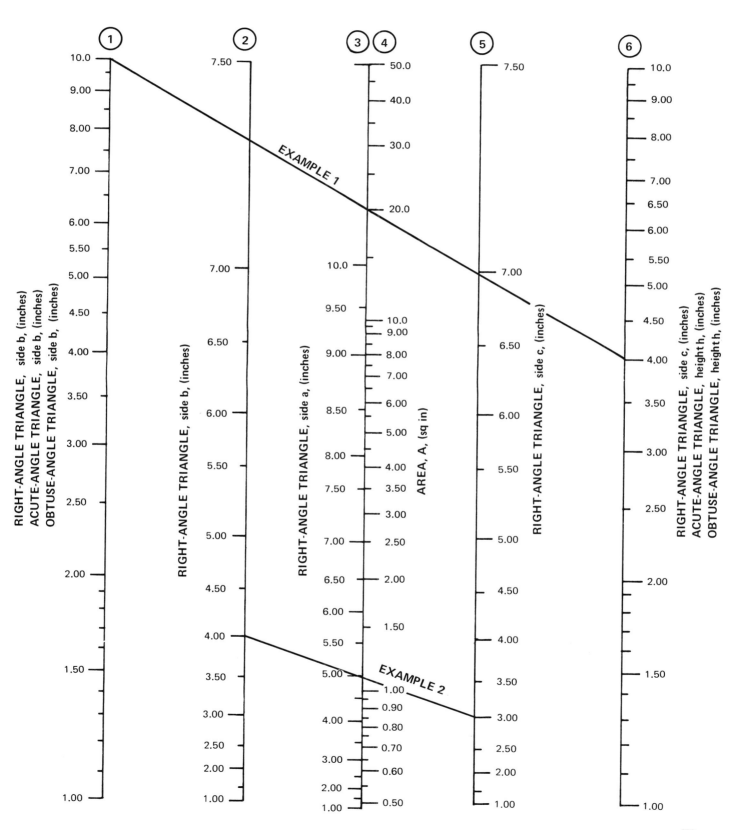

Area of spandrel

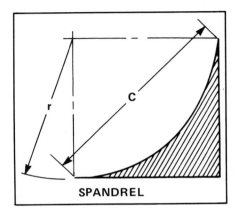

SPANDREL

Introduction.
To determine the area of a spandrel or fillet the following equations can be used:

$$A = 0.215r^2$$
$$A = 0.1075C^2$$
$$A = r^2 - \frac{\pi r^2}{4}$$

Nomenclature:
A = Area, sq in
r = Radius, inches
C = Chord, inches

Nomogram.
A nomogram can be used to expedite the solution of the above equations.

Example 1:
Given a spandrel that has a radius of 0.70 inches, determine its area.

Solution 1:
Construct a line from 0.70 on scale 3 to $A = 0.215r^2$ on scale 1 and where this line intersects scale 2 read the answer of 0.10 sq in.

Example 2:
Given a spandrel that has a chord of 0.22 inches, determine its area.

Solution 2:
Construct a line from $A = 0.1075C^2$ on scale 1 to 0.22 on scale 3 and where this line intersects scale 2 read the answer of 0.005 sq in.

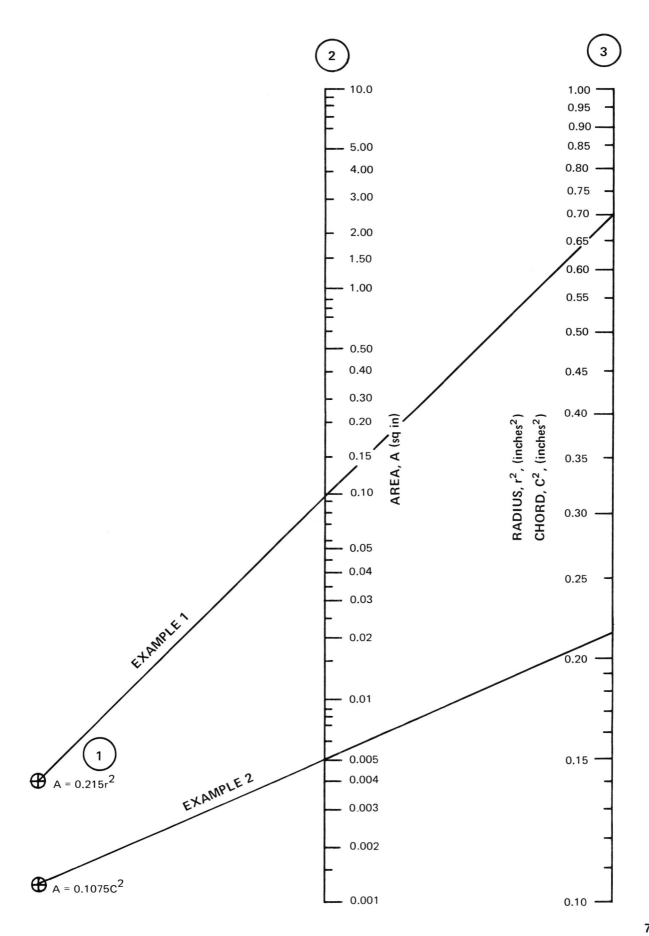

Area of spherical surface

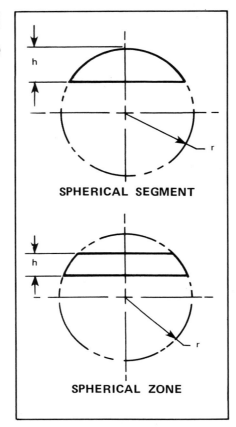

SPHERICAL SEGMENT

SPHERICAL ZONE

Introduction.
To determine the area of the spherical surface of a spherical segment and a spherical zone, the following equations can be used.

$$A = 2\pi rh$$
$$A = 6.2832rh$$

Nomenclature:
A = Area, sq in.
r = Radius, inches
h = Height, inches

Nomogram.
A nomogram can be used to expedite the solution of the above equations.

Example 1:
Given a spherical segment that has a radius of 5.00 inches and height of 3.00 inches, determine the area of its spherical surface.

Solution 1:
Construct a line from 5.00 on scale 3 to A = 6.2832rh on scale 1 and where this line intersects the reference line label point A. From point A construct a line to 3.00 on scale 4 and where this line intersects scale 2 read the answer of 94.0 sq. in.

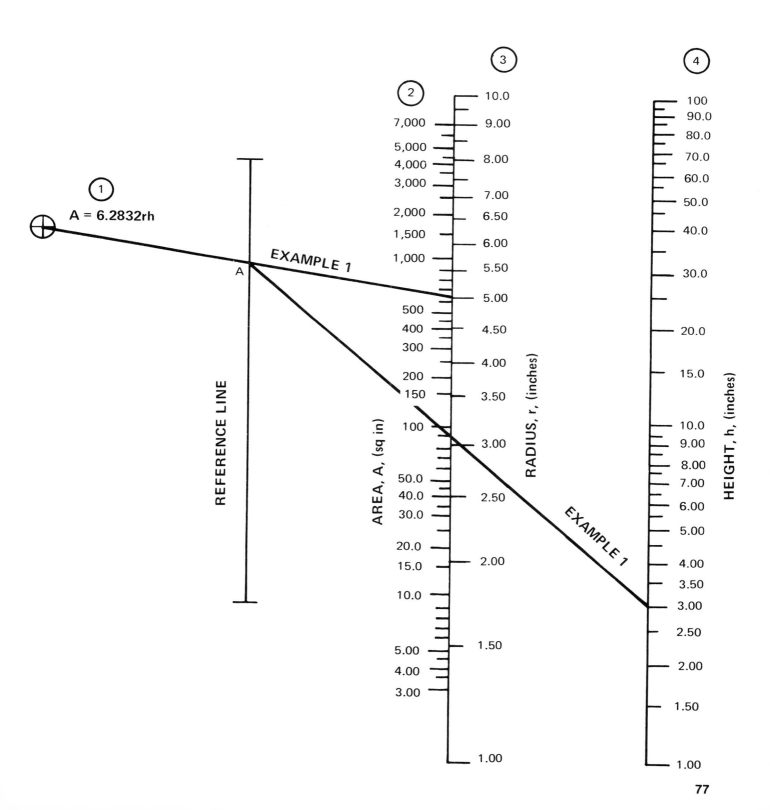

① A = 6.2832rh

② AREA, A, (sq in)

③ RADIUS, r, (inches)

④ HEIGHT, h, (inches)

REFERENCE LINE

EXAMPLE 1

EXAMPLE 1

Area of square, rectangle and parallelogram

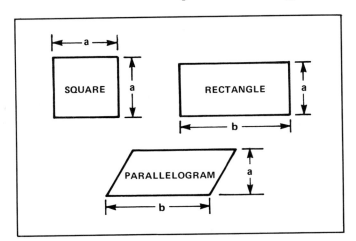

Introduction.

To determine the area of a square, rectangle and parallelogram the following equations can be used:

$$\text{Square: } A = a^2$$
$$\text{Rectangle: } A = ab$$
$$a = \frac{A}{b}$$
$$b = \frac{A}{a}$$
$$\text{Parallelogram: } A = ab$$
$$a = \frac{A}{b}$$
$$b = \frac{A}{a}$$

Nomenclature:

A = Area, sq in
a = side a, inches
b = side b, inches

Nomogram.

A nomogram can be used to expedite the solution of the above equations. When using the nomogram, the following procedure should be followed:

Square: To find the area, use scale 1 and read the answer on scale 2.

Rectangle: To find the area, use scales 1 and 4 and read the answer on scale 3.

Parallelogram: To find the area, use scales 1 and 4 and read the answer on scale 3.

Example 1:

Given one side of a square as 5.00 inches, determine its area.

Solution 1:

Locate 5.00 on scale 1 and read the answer of 25.0 sq in on scale 2.

Example 2:

Given a rectangle that has side a of 2.00 inches and side b of 3.00 inches, determine its area.

Solution 2:

On the nomogram construct a line from 2.00 on scale 1 to 3.00 on scale 4 and where this line intersects scale 3 read the answer of 6.00 sq in.

Example 3:

Given a parallelogram that has an area of 60.0 sq in and side a of 6.00 inches, determine side b.

Solution 3:

On the nomogram construct a line from 6.00 on scale 1 to 60.0 on scale 3 and continue this line until it intersects scale 4. At this intersection read the answer of 10.0 inches.

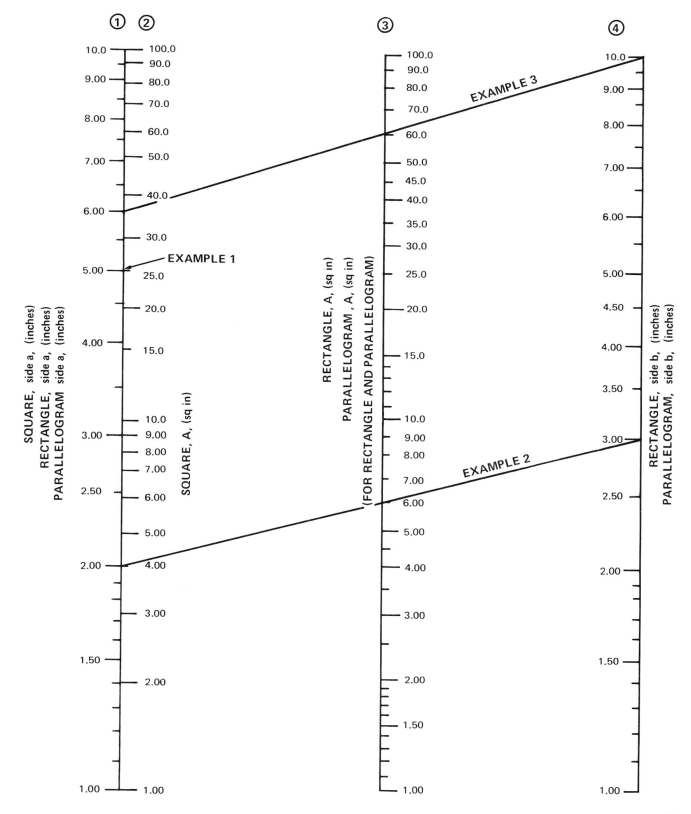

Area of trapezoid

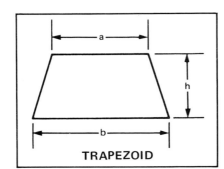

TRAPEZOID

To determine the area of a trapezoid, the following equation can be used:

$$A = \frac{(a + b)h}{2}$$

Nomenclature:
A = Area, sq in
a = side a, inches
b = side b, inches
h = height, inches

Nomogram.
A nomogram can be used to expedite the solution of the above equation.

Example 1:
Given a trapezoid that has side a = 20.0 inches, side b = 30.0 inches and height h = 10.0 inches, determine the area.

Solution 1:
On the nomogram construct a line from 20.0 on the side a scale to 30.0 on the side b scale and where this line intersects the reference line, label point A. From this point A, construct a horizontal line to 10.0 on the height h scale and mark this intersection, point B. From this point B, construct a vertical line down to the Area scale and read the answer of 250 sq in.

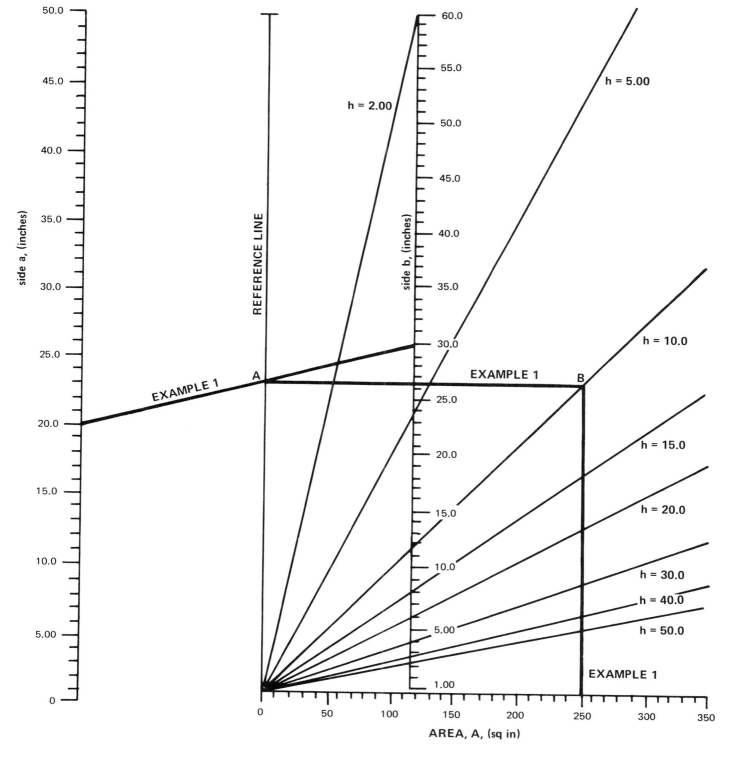

Volume of cone

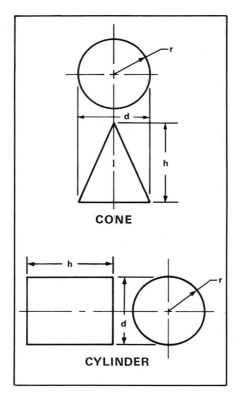

CONE

CYLINDER

Introduction.

To determine the volume of a cone or cylinder the following equations can be used:

$$\text{Cone: } V = 1.0472r^2h$$
$$V = 0.2618d^2h$$
$$\text{Cylinder: } V = 3.1416r^2h$$
$$V = 0.7834d^2h$$

Nomenclature:

V = Volume, cu in
r = Radius, inches
d = Diameter, inches
h = Height, inches

Nomogram.

A nomogram can be used to expedite the solution of the above equations.

Example 1:

Given a cone that has a diameter of 1.50 inches and height of 5.00 inches, determine its volume.

Solution 1:

Construct a line from 1.50 on scale 1 to 5.00 on scale 5 and where this line intersects scale 4 read the answer of 2.95 cu in.

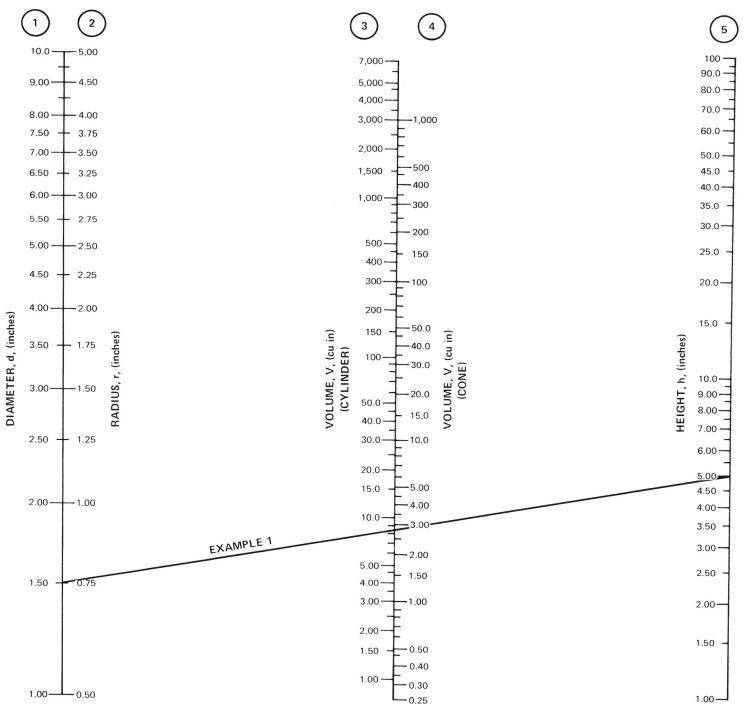

Volume of cube

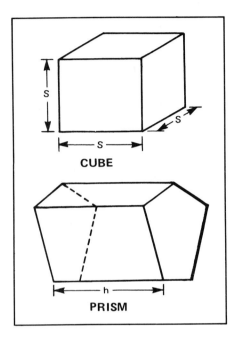

CUBE

PRISM

Introduction.
To determine the volume of a cube or prism the following equations can be used:

Cube: $V = S^3$
$S = \sqrt[3]{V}$

Prism: $V = Ah$

Nomenclature:
 V = Volume, cu in
 S = Side of cube, inches
 h = Height of prism, inches
 A = Area of end surface, sq in

Nomogram.
 A nomogram can be used to expedite the solution of the above equations. When using the nomogram the following procedure should be followed:
 To find the volume of a cube use scales 1 and 2.
 To find the volume of a prism use scales 1, 3 and 4.

Example 1:
 Given a cube that has a side of 8.00 inches, determine its volume.

Solution 1:
 Locate 8.00 on scale 1 and read the answer of 512 cu in on scale 2.

Example 2:
 Given a prism that has a height of 5.00 inches and volume of 12.5 cu in, determine the area of the end surface.

Solution 2:
 Construct a line from 5.00 on scale 1 to 12.5 on scale 3 and continue this line until it intersects scale 4 and read the answer of 2.50 sq in.

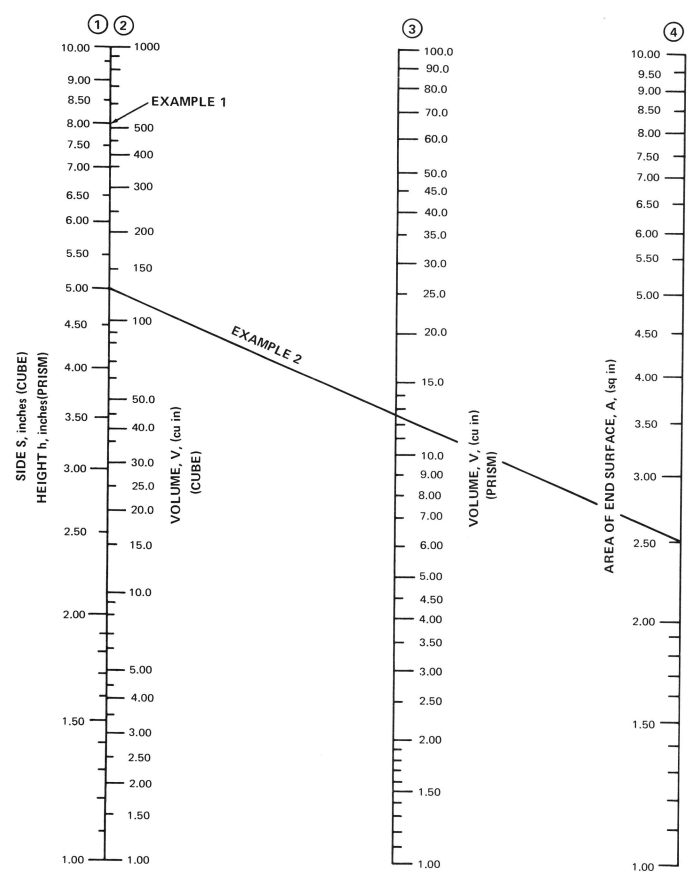

Volume of ellipsoid of revolution (case 1)

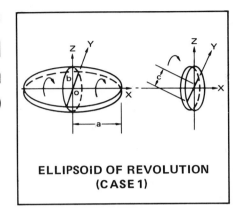

**ELLIPSOID OF REVOLUTION
(CASE 1)**

Introduction.

To determine the volume of an ellipsoid of revolution or spheroid the following equations can be used:

$$V = \frac{4\pi}{3} abc$$
$$V = 4.1888abc$$

Nomenclature:
 V = Volume, cu in
 a = X-axis, inches
 b = Y-axis, inches
 c = Z-axis, inches

Nomogram.
 A nomogram can be used to expedite the solution of the above equations.

Example 1:
 Given an ellipsoid of revolution that has an X-axis of 5.00 inches, Y-axis of 4.00 inches and Z-axis of 3.00 inches, determine its volume.

Solution 1:
 Construct a line from 5.00 on scale 1 to 4.00 on scale 2 and where this line intersects the reference line, label point A. Construct a line from point A to 3.00 on scale 4 and where this line intersects scale 3 read the answer of 251 cu in.

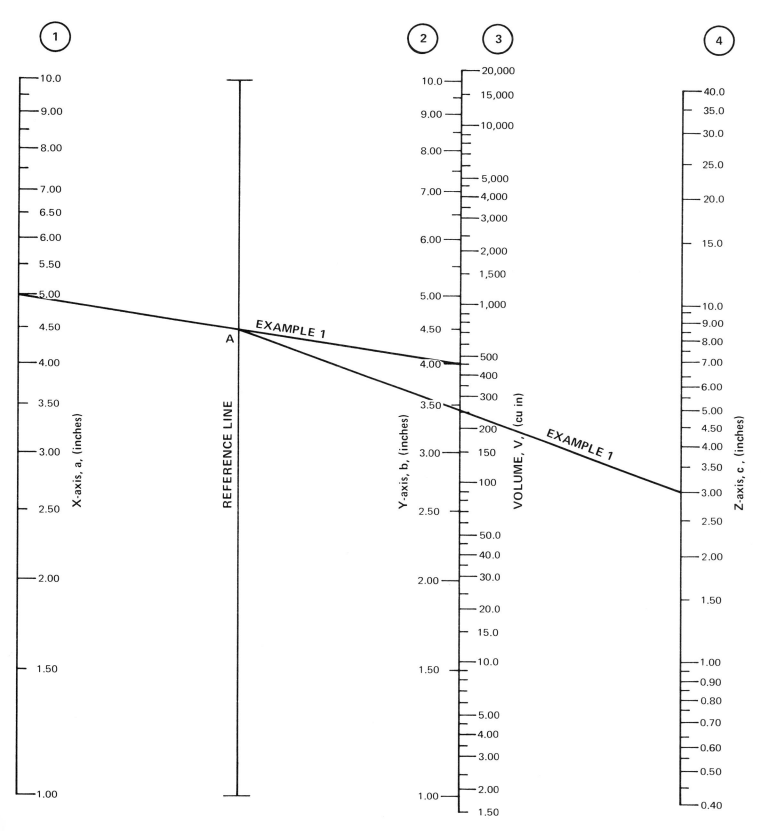

Volume of ellipsoid of revolution (case 2)

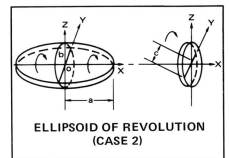

ELLIPSOID OF REVOLUTION (CASE 2)

Introduction.
To determine the volume of an ellipsoid of revolution or spheroid the following equations can be used:

$$V = \frac{4\pi abc}{3}$$

$$V = 4.1888ab^2$$

Nomenclature:
V = Volume, cu in
a = X-axis, inches
b = c = Y-axis, inches

Nomogram.
A nomogram can be used to expedite the solution of the above equations.

Example 1:
Given an ellipsoid of revolution that has a X-axis of 5.00 inches and Y-axis of 1.50 inches, determine its volume.

Solution 1:
Construct a line from 5.00 on scale 1 to 1.50 on scale 3 and where this line intersects scale 2 read the answer of 47.0 cu in.

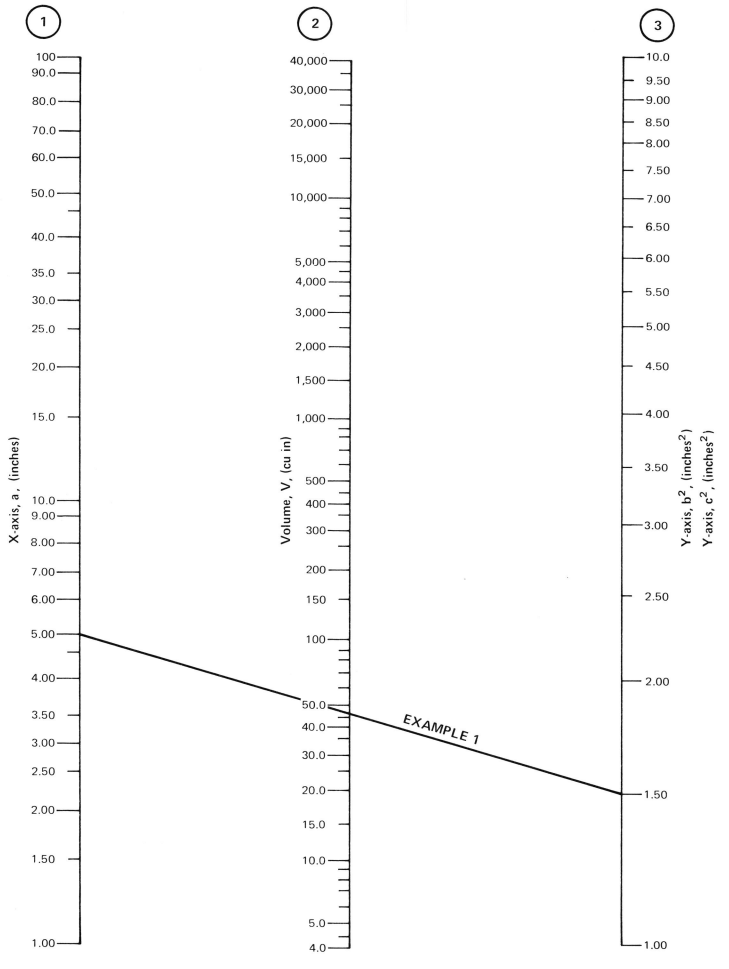

Volume of paraboloid

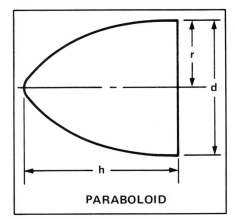

PARABOLOID

Introduction.
To determine the volume of a paraboloid the following equations can be used:

$$V = \frac{1}{2}\pi r^2 h$$

$$V = 0.3927 d^2 h$$

Nomenclature:
V = Volume, cu in
d = Diameter, inches
r = Radius, inches
h = Height, inches

Nomogram.
A nomogram can be used to expedite the solution of the above equations.

Example 1:
Given a paraboloid that has a diameter of 5.00 inches and height of 15.0 inches, determine its volume.

Solution 1:
Construct a line from 15.0 on scale 1 to 5.00 on scale 3 and where this line intersects scale 2 read the answer of 147 cu in.

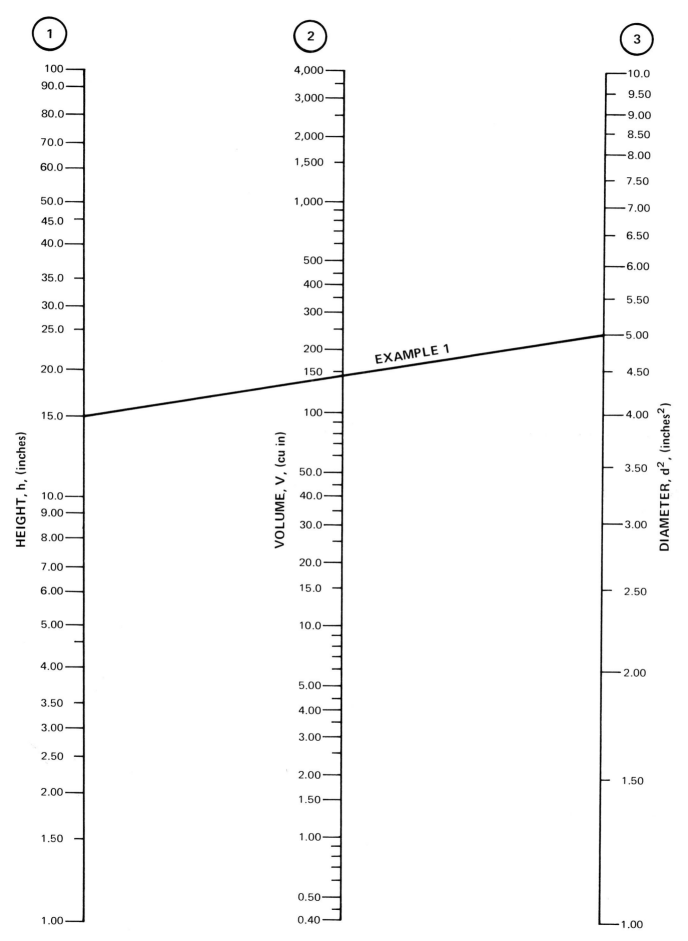

EXAMPLE 1

HEIGHT, h, (inches)

VOLUME, V, (cu in)

DIAMETER, d^2, (inches^2)

Volume of pyramid

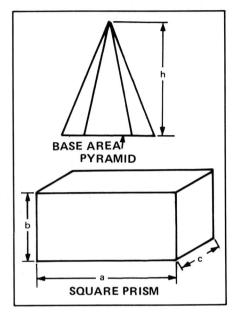

BASE AREA
PYRAMID

SQUARE PRISM

Introduction.
To determine the volume of a pyramid or a square prism, the following equations can be used:

$$\text{Pyramid: } V = \frac{1}{3}hA$$

$$\text{Square Prism: } V = abc$$

Nomenclature:
V = Volume, cu in
Side a = Side of square prism, inches
Side b = Side of square prism, inches
Side c = Side of square prism, inches
h = Height, inches
A = Area of base, sq in

Nomogram.
A nomogram can be used to expedite the solution of the above equations.

Example 1:
Given a pyramid that has a height of 9.0 inches and area of base of 6.0 sq in, determine its volume.

Solution 1:
On the nomogram construct a line from 9.0 on the height scale to 6.0 on the Area of base scale. From this point construct a horizontal line (to the left) until it intersects the reference line. From this point construct a vertical line (downward) until it intersects the Volume scale and read the answer of 18 cu in.

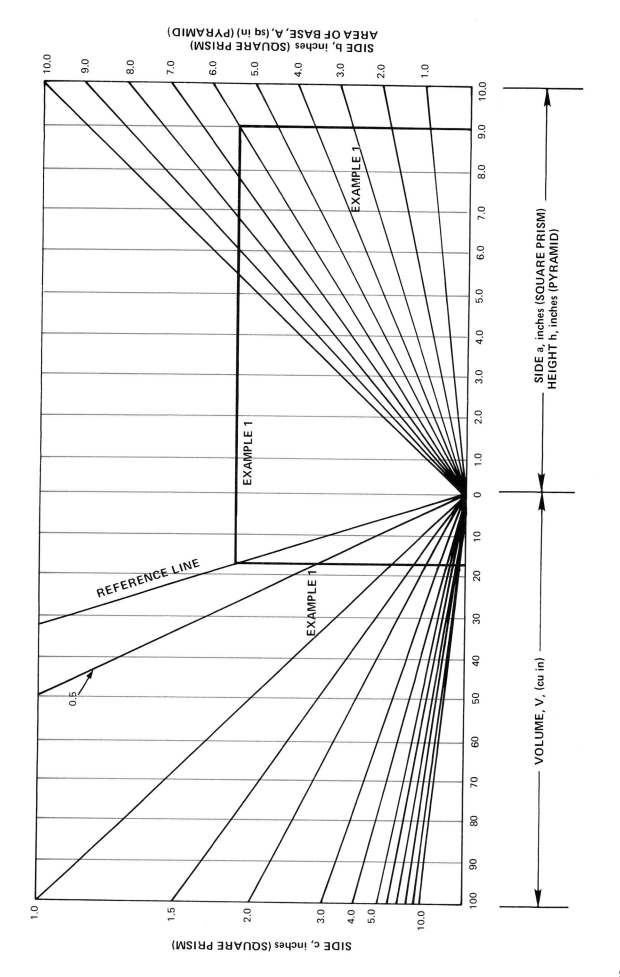

Volume of rings

Introduction.

In engineering design and production situations, the engineer is frequently faced with the problem of finding the volume of a ring. Sometimes, because of the cross-sectional configuration being other than rectangular or circular, he is usually compelled to resort to integration equations or an approximation.

Eq. 1 was used in the construction of the following table that gives the various shapes usually encountered and the relevant equations.

$$V = A2\pi C_x \tag{1}$$

Nomenclature:

A = Area of cross-section of ring, sq in

C_x = Distance of center of gravity of section from the axis of the ring, inch

V = Volume of ring, cu in

Derivation of a typical equation.

The following is the derivation of the equation for the volume of a ring (Type 1) in the table.

$$V = A2\pi C_x$$

$$V = \frac{ab2\pi}{2}\left(R + \frac{b}{3}\right)$$

$$V = \frac{ab2\pi R}{2} + \frac{ab2\pi b}{6}$$

$$V = \pi Rab + \frac{ab^2\pi}{3}$$

This is the equation found in the volume column for a Type 1 problem.

Example 1:

Determine the volume of a Type 1 ring that has $a = 1$ inch, $b = 2$ inches and $R = 3$ inches.

Solution 1:

Substituting into the equation for finding the volume of a Type 1 ring:

$$V = \pi Rab + \frac{ab^2\pi}{3}$$

$$V = (3.1416)(3)(1)(2) + \frac{(1)(2^2)(3.1416)}{3}$$

$$V = 18.84 + 4.18$$
$$V = 23.02 \text{ cu in}$$

	TYPE	A	C_x	V
1		$\dfrac{ab}{2}$	$R + \dfrac{b}{3}$	$\pi Rab + \dfrac{ab^2\pi}{3}$
2			$R - \dfrac{b}{3}$	$\pi Rab - \dfrac{ab^2\pi}{3}$
3		$\dfrac{\pi r^2}{4}$	$R + \dfrac{4r}{3\pi} = R + 0.4244r$	$\dfrac{\pi^2 r^2 R}{2} + \dfrac{2\pi r^3}{3}$ $= 4.9348r^2R + 2.0944r^3$
4			$R - \dfrac{4r}{3\pi} = R - 0.4244r$	$\dfrac{\pi^2 r^2 R}{2} - \dfrac{2\pi r^3}{3}$ $= 4.9348r^2R - 2.0944r^3$
5		$0.2146r^2$	$R + 0.2218r$	$1.3486Rr^2 + 0.3011r^3$
6			$R - 0.2218r$	$1.3486Rr^2 - 0.3011r^3$
7	QUADRANT OF AN ELLIPSE	$\dfrac{\pi}{4}ab$ $= 0.7854ab$	$R + \dfrac{4a}{3\pi} = R + 0.4244a$	$\dfrac{\pi^2 abR}{2} + \dfrac{2\pi a^2 b}{3}$ $= 4.9348abR + 2.0944a^2b$
8			$R - \dfrac{4a}{3\pi} = R - 0.4244a$	$\dfrac{\pi^2 abR}{2} - \dfrac{2\pi a^2 b}{3}$ $= 4.9348abR - 2.0944a^2b$
9	QUADRANT OF A QUADRATIC PARABOLA	$\dfrac{2}{3}Wh$	$\left(R + \dfrac{2h}{5}\right) = (R + 0.4h)$	$\dfrac{4\pi Rwh}{3} + \dfrac{\pi wh^2}{3} \times 1.6$ $= 4.1888Rwh + 1.6755wh^2$
10			$\left(R - \dfrac{2h}{5}\right) = (R - 0.4h)$	$\dfrac{4\pi Rwh}{3} - \dfrac{\pi wh^2}{3} \times 1.6$ $= 4.1888Rwh - 1.6755wh^2$

Volume of sphere

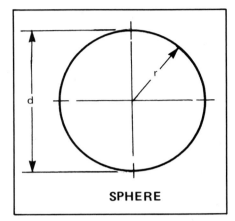

SPHERE

Introduction.

To determine the volume and area of surface of a sphere the following equations can be used:

$$V = \frac{4\pi r^3}{3}$$

$$V = \frac{\pi d^3}{6}$$

$$V = 4.1888 r^3$$

$$V = 0.5236 d^3$$

$$A = 4\pi r^2$$

$$A = \pi d^2$$

$$A = 12.5664 r^2$$

$$A = 3.1416 d^2$$

Nomenclature:

V = Volume, cu in
A = Area of surface, sq in
r = Radius, inches
d = Diameter, inches

Nomogram.

A nomogram can be used to expedite the solution of the above equations. When using the nomogram the following procedure should be followed:

To find the Volume use scales 1, 2 and 4.

To find the Area use scales 1, 3 and 5.

Example 1:

Given a sphere that has a radius of 3.00 inches, determine its volume.

Solution 1:

Construct a line from 3.00 on scale 4 to $V = 4.1888 r^3$ on scale 1 and where this line intersects scale 2 read the answer of 113 cu in.

Example 2:

Given a sphere that has a diameter of 1.50 inches, determine the area of its surface.

Solution 2:

Construct a line from 1.50 on scale 5 to $A = 3.1416 d^2$ on scale 1 and where this line intersects scale 3 read the answer of 7.07 sq in.

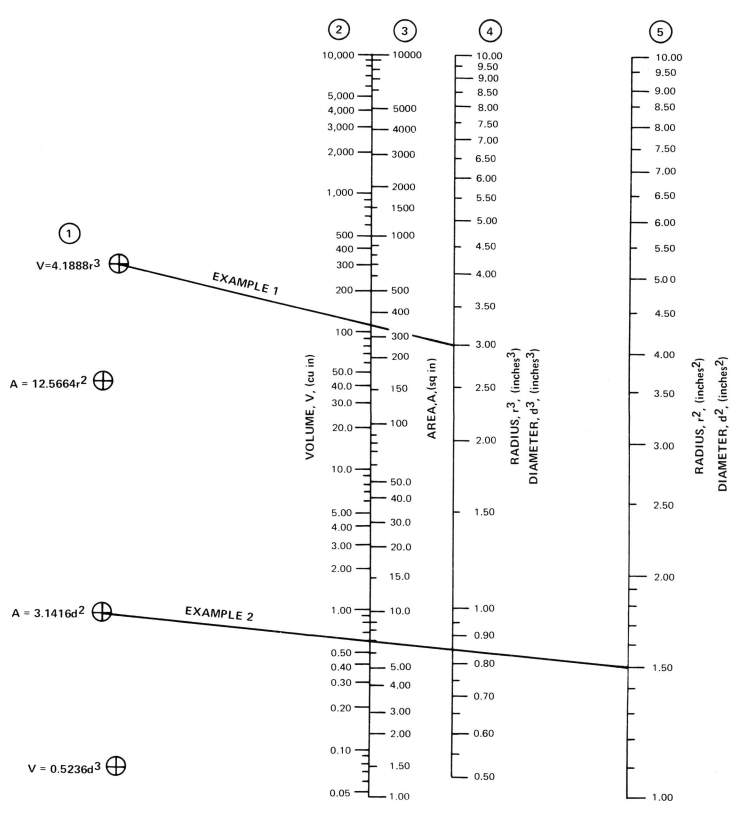

① V=4.1888r³ ⊕

A = 12.5664r² ⊕

A = 3.1416d² ⊕

V = 0.5236d³ ⊕

EXAMPLE 1

EXAMPLE 2

② 10,000 — 5,000 — 4,000 — 3,000 — 2,000 — 1,000 — 500 — 400 — 300 — 200 — 100 — 50.0 — 40.0 — 30.0 — 20.0 — 10.0 — 5.00 — 4.00 — 3.00 — 2.00 — 1.00 — 0.50 — 0.40 — 0.30 — 0.20 — 0.10 — 0.05

VOLUME, V, (cu in)

③ 10000 — 5000 — 4000 — 3000 — 2000 — 1500 — 1000 — 500 — 400 — 300 — 200 — 150 — 100 — 50.0 — 40.0 — 30.0 — 20.0 — 15.0 — 10.0 — 5.00 — 4.00 — 3.00 — 2.00 — 1.00

AREA, A, (sq in)

④ 10.00 — 9.50 — 9.00 — 8.50 — 8.00 — 7.50 — 7.00 — 6.50 — 6.00 — 5.50 — 5.00 — 4.50 — 4.00 — 3.50 — 3.00 — 2.50 — 2.00 — 1.50 — 1.00 — 0.90 — 0.80 — 0.70 — 0.60 — 0.50

RADIUS, r^3, (inches³)
DIAMETER, d^3, (inches³)

⑤ 10.00 — 9.50 — 9.00 — 8.50 — 8.00 — 7.50 — 7.00 — 6.50 — 6.00 — 5.50 — 5.00 — 4.50 — 4.00 — 3.50 — 3.00 — 2.50 — 2.00 — 1.50 — 1.00

RADIUS, r^2, (inches²)
DIAMETER, d^2, (inches²)

Volume of spherical sector

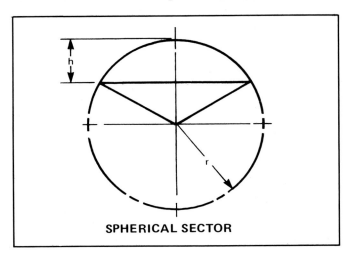

SPHERICAL SECTOR

Introduction.
To determine the volume of a spherical sector the following equations can be used:

$$V = \frac{2\pi r^2 h}{3}$$

$$V = 2.0944 r^2 h$$

Nomenclature:
  V = Volume, cu in
  r = Radius, inches
  h = Height, inches

Nomogram.
  A nomogram can be used to expedite the solution of the above equations.

Example 1:
  Given a spherical sector that has a radius of 3.00 inches and height of 1.00 inch, determine its volume.

Solution 1:
  Construct a line from 1.00 on scale 1 to 3.00 on scale 3 and where this line intersects scale 2 read the answer of 19.0 cu in.

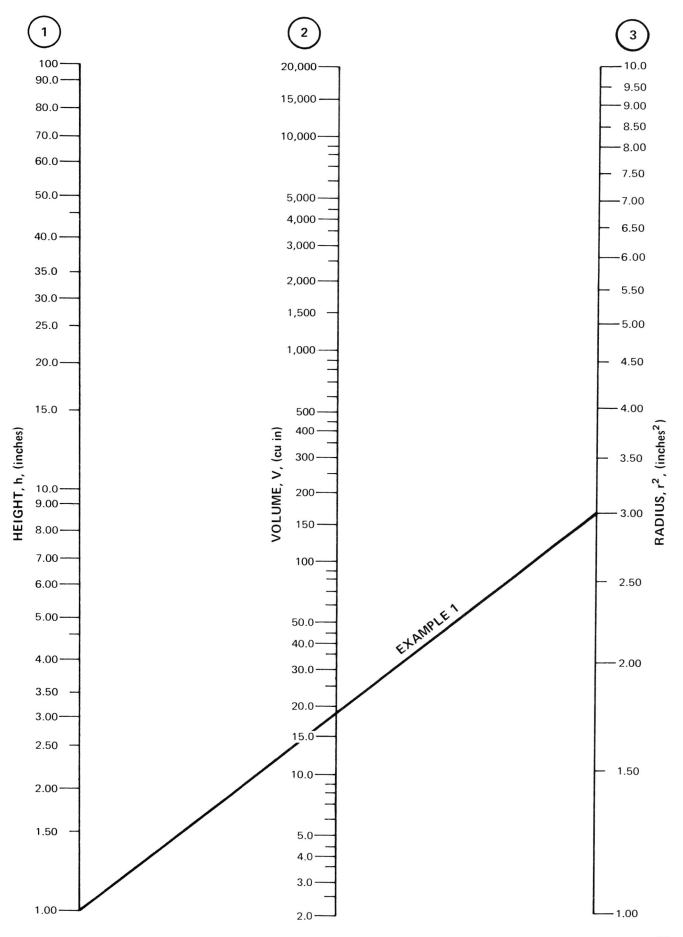

Volume of spherical wedge

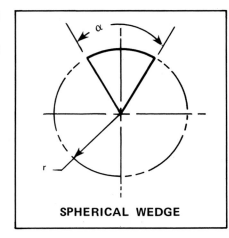

SPHERICAL WEDGE

Introduction.
To determine the volume and area of a spherical wedge the following equations can be used:

$$V = \frac{\alpha}{360}\left(\frac{4\pi r^3}{3}\right)$$

$$V = 0.0116\,\alpha r^3$$

$$A = \frac{\alpha}{360}\left(4\pi r^2\right)$$

$$A = 0.0349\,\alpha r^2$$

Nomenclature:
V = Volume, cu in
A = Area, sq in
r = Radius, inches
α = Central angle, deg

Nomogram.
A nomogram can be used to expedite the solution of the above equations.

Example 1:
Given a spherical wedge that has a radius of 2.00 inches and central angle of 45°, determine its volume.

Solution 1:
Construct a line from 45 on scale 1 to 2.00 on scale 5 and where this line intersects scale 4 read the answer of 4.18 cu in.

Example 2:
Given a spherical wedge that has a radius of 1.00 inch and central angle of 10°, determine its area.

Solution 2:
Construct a line from 10 on scale 2 to 1.00 on scale 5 and where this line intersects scale 3 read the answer of 0.35 sq in.

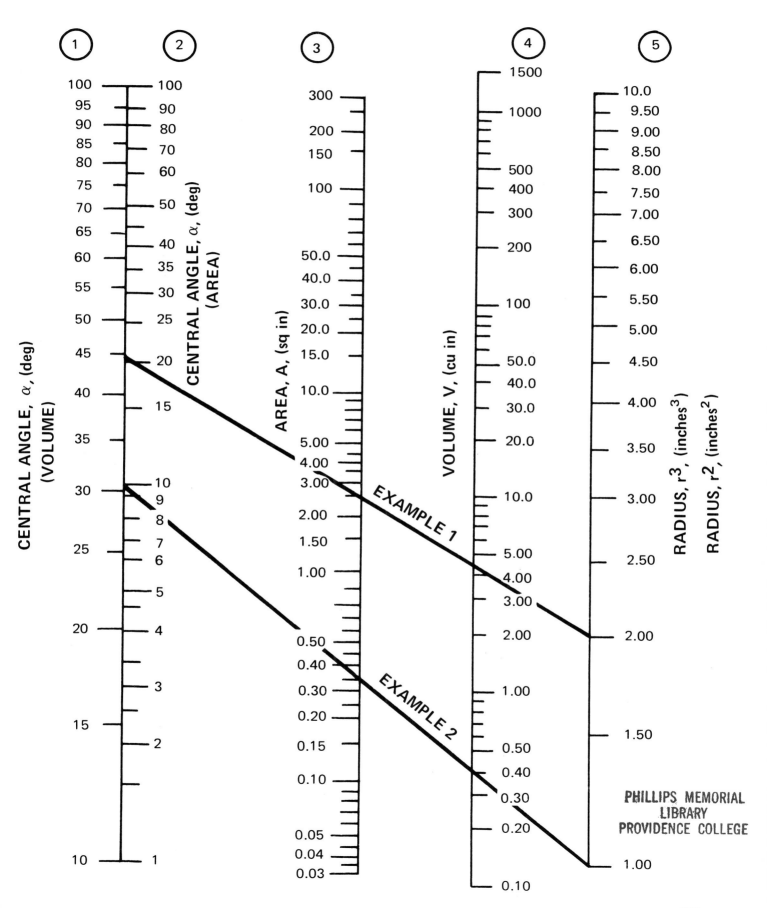

CENTRAL ANGLE, α, (deg) (VOLUME)

CENTRAL ANGLE, α, (deg) (AREA)

AREA, A, (sq in)

VOLUME, V, (cu in)

RADIUS, r^3, (inches^3)

RADIUS, r^2, (inches^2)

EXAMPLE 1

EXAMPLE 2

Volume of solids

Introduction.
The prismoidal equation provides a convenient means for finding the volume of many solids widely used in engineering. In fact, an equation for the volume usually can be derived more quickly from this equation than reference can be made to a handbook. The prismoidal equation applies to solids enclosed by plane surfaces and to certain bodies with curved surfaces.

$$V = \frac{h}{6}(A_{base} + 4A_m + A_{top}) \qquad (1)$$

Nomenclature:

V = Volume of solid, cu in
A_{base} = Area of the bottom surface of the body, sq in
A_{top} = Area of the top surface of the body, sq in
A_m = Area of the cross-section halfway between the top and bottom, sq in
h = Perpendicular distance between parallel surfaces, inches

Notes.
A_m is not $1/2 (A_{base} + A_{top})$ and Eq 1 has been solved for the following solids.

Example 1:
Given a sphere that has a radius of 3.0 inches, determine its volume.

Solution 1:
Going to the sphere portion of the article, we arrive at the following equation:

$$V = \frac{4}{3}\pi r^3$$

Substituting into this equation the value of the radius:

$$V = \frac{4}{3}(3.1416)(3.0)^3$$

$$V = \frac{4}{3}(3.1416)(27.0)$$

$$V = \frac{4}{3}(84.8)$$

$$V = 113.1 \text{ cu in}$$

FOUR-SIDED SOLID

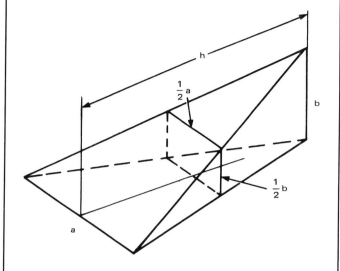

$$V = \frac{h}{6}\left(0 + 4 \times \frac{a}{2} \times \frac{b}{2} + 0\right)$$

$$V = \frac{hab}{6}$$

WEDGE

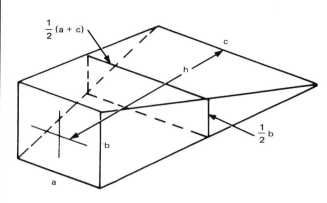

$$V = \frac{h}{6}\left(ab + 4 \times \frac{b}{2} \times \frac{a+c}{2} + 0\right)$$

$$V = \frac{hb}{6}(2a + c)$$

SPHERE

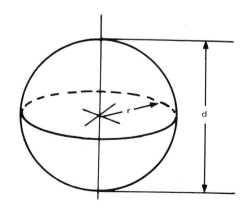

$$V = \frac{d}{6}\left(0 + 4 \times \frac{\pi d^2}{4} + 0\right)$$

$$V = \frac{\pi d^3}{6}$$

$$V = \frac{4}{3}\pi r^3$$

SPHERICAL SEGMENT

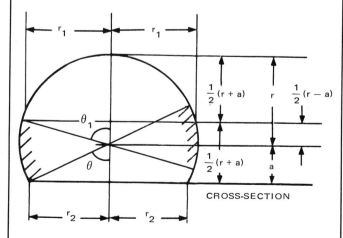

CROSS-SECTION

$$\cos\theta_1 = \frac{r-a}{2r} = \frac{1}{2}\left(1 - \frac{a}{r}\right) \qquad r_1 = r\,\text{SIN}\,\theta_1$$

$$\cos\theta = \frac{a}{r} \qquad\qquad\qquad r_2 = r\,\text{SIN}\,\theta$$

$$V = \frac{(r+a)}{6}\left(\pi r_2^2 + 4\pi r_1^2 + 0\right)$$

$$V = \frac{\pi(r+a)}{6}\left(4r_1^2 + r_2^2\right)$$

PYRAMID

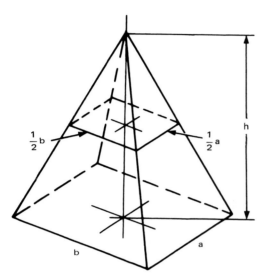

$\frac{1}{2}b$ $\frac{1}{2}a$

$V = \frac{h}{6}(ab + 4 \times \frac{a}{2} \times \frac{b}{2} + 0)$

$V = \frac{h}{6} \times 2ab$

$V = \frac{hab}{3}$

CONE

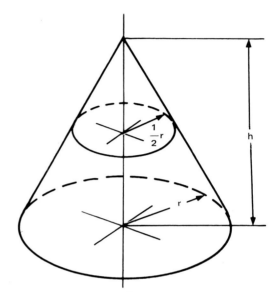

$\frac{1}{2}r$

$V = \frac{h}{6}[\pi r^2 + 4\pi (\frac{r}{2})^2 + 0]$

$V = \frac{\pi h}{6} \times 2r^2$

$V = \frac{\pi h r^2}{3}$

FRUSTUM OF PYRAMID

b_1 a_1

$\frac{b + b_1}{2}$ $\frac{a + a_1}{2}$

$V = \frac{h}{6}\left[ab + 4(\frac{a + a_1}{2})(\frac{b + b_1}{2}) + a_1 b_1\right]$

$V = \frac{h}{6}(2ab + 2a_1 b_1 + ab_1 + a_1 b)$

FRUSTUM OF CONE

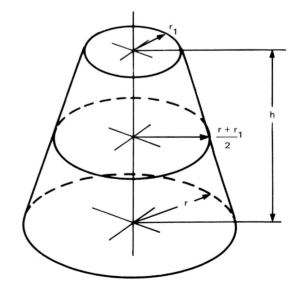

r_1

$\frac{r + r_1}{2}$

$V = \frac{h}{6}[\pi r^2 + 4\pi (\frac{r + r_1}{2})^2 + \pi r_1^2]$

$V = \frac{\pi h}{3}(r^2 + r_1 r + r_1^2)$

Cones with rounded tips

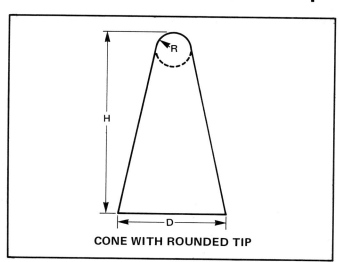

CONE WITH ROUNDED TIP

Introduction.
It is more practical to use a cone with a rounded point than one with a sharp tip. The equations for the volume and surface area of such an arrangement are not readily available in handbooks. If the rounded portion is spherical, the following method will produce fairly accurate answers.

Nomenclature:

D = Diameter of the base, inches
R = Radius of the spherical portion, inches
H = Total height, inches
$d = \dfrac{D}{H}$, dimensionless
$r = \dfrac{R}{H}$, dimensionless

The simplified equations for the volume and area are:

$V = H^3v$ v is obtained from Nomogram 1   (1)
$A = H^2a$ a is obtained from Nomogram 2   (2)

Example 1:
Given a cone with a rounded point that has H = 3.0 inches, D = 2.0 inches and R = 0.5 inches, determine the volume and surface area of this cone.

Solution 1:
Calculate the following:

$$r = \frac{R}{H}$$
$$r = \frac{0.5}{3.0}$$
$$r = 0.167$$
$$d = \frac{D}{H}$$
$$d = \frac{2.0}{3.0}$$
$$d = 0.667$$

Proceed to Nomogram 1 and let r = 0.167 and d = 0.667 and at their intersection point read the answer of v = 0.180.
Substituting back into Eq. 1:

$$V = H^3v$$
$$V = (3.0^3)(0.180)$$
$$V = 4.86 \text{ cu in}$$

Next, proceed to Nomogram 2 and let r = 0.167 and d = 0.667 and at their intersection point read the answer of a = 1.52.
Substituting back into Eq. 2:

$$A = H^2a$$
$$A = (3.0^2)(1.52)$$
$$A = 13.68 \text{ sq in}$$

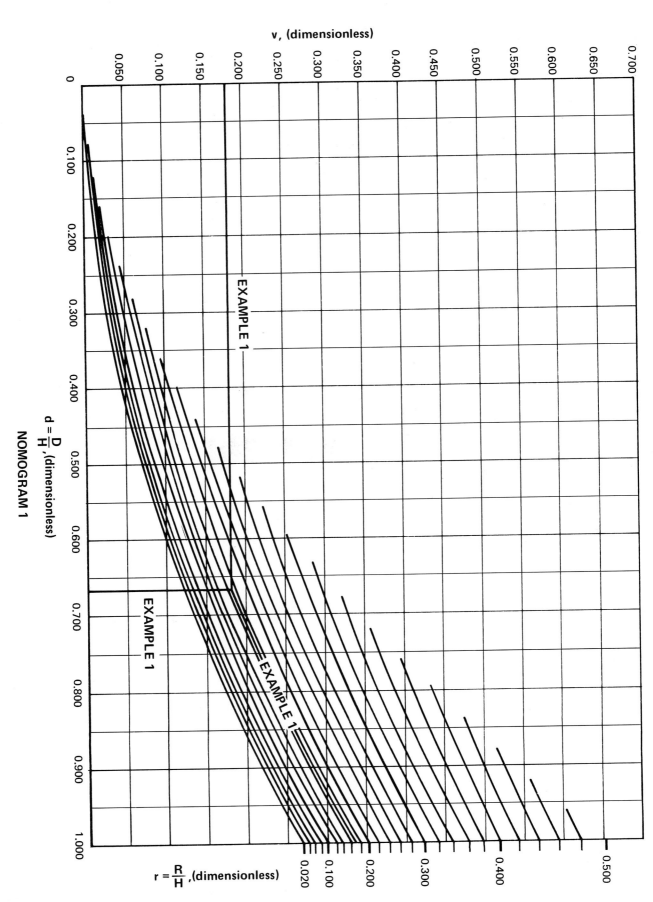

v, (dimensionless)

d = D/H , (dimensionless)

NOMOGRAM 1

r = R/H , (dimensionless)

EXAMPLE 1

EXAMPLE 1

EXAMPLE 1

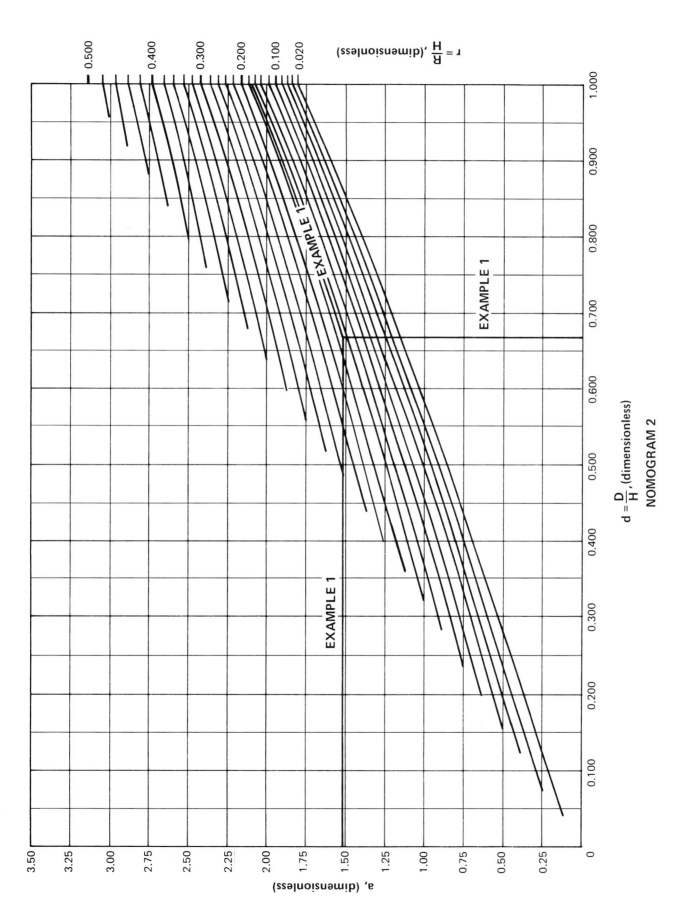

NOMOGRAM 2

Angles of regular polygon

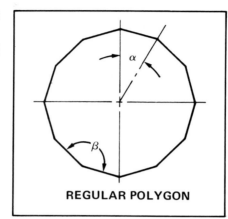

REGULAR POLYGON

Introduction
To determine the angles α and β in a regular polygon the following equations can be used:

$$\alpha = \frac{360°}{n}$$

$$\beta = 180° - \alpha$$

Nomenclature:
α = Acute angle, deg
β = Obtuse angle, deg
n = Number of sides

Nomogram.
A nomogram can be used to expedite the solution of the above equations. When using the nomogram, the following procedure should be followed:
To find α use scales 1, 4 and 6.
To find β use scales 2, 3 and 5.

Example 1:
Given a regular polygon that has 12 sides, determine the acute angle α.

Solution 1:
Construct a line from 12 on scale 6 to $\alpha = \dfrac{360°}{n}$ on scale 1 and where this line intersects scale 4 read the answer of 30°.

Example 2:
Using the acute angle from Example 1, determine the obtuse angle β.

Solution 2:
Construct a line from 30 on scale 2 to $\beta = 180° - \alpha$ on scale 3 and continue this line until it intersects scale 5. At this intersection point read the answer of $\beta = 150°$.

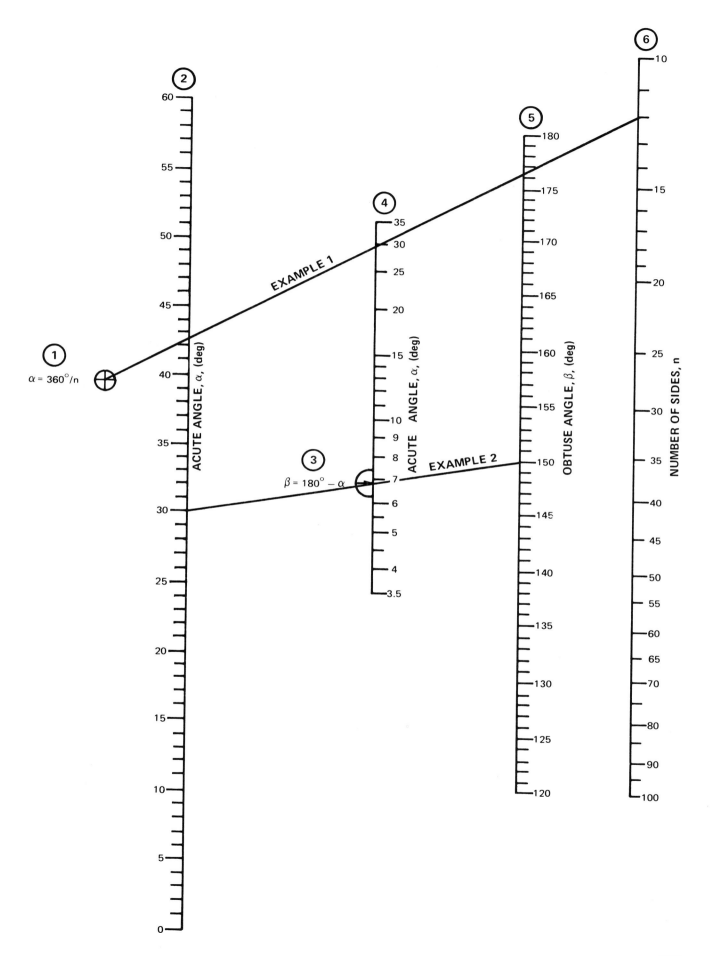

① $\alpha = 360°/n$

② ACUTE ANGLE, α, (deg)

③ $\beta = 180° - \alpha$

④ ACUTE ANGLE, α, (deg)

⑤ OBTUSE ANGLE, β, (deg)

⑥ NUMBER OF SIDES, n

EXAMPLE 1

EXAMPLE 2

Nomogram for law of sines

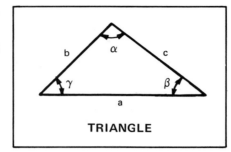

TRIANGLE

Introduction.

Frequently, triangles are solved by making use of the law of sines:

$$\frac{a}{\sin \alpha} = \frac{b}{\sin \beta} = \frac{c}{\sin \gamma} \qquad (1)$$

Using trigonometric identities and rearranging terms, the following equation is arrived at:

$$a = \frac{c \sin \alpha}{\sin(\alpha + \beta)} \qquad (2)$$

The following nomogram provides a quick solution to Eq. 2.

Example 1:

Given a triangle that has c = 10 inches, α = 50 deg and β = 30 deg, determine the length of side a of the triangle.

Solution 1:

Construct a line from c = 10 to α = 50 deg and mark the intersection point of this line with β = 30 deg. The particular curve of the a scale that intersects this point is found by interpolation to have a value of a = 7.8 inches.

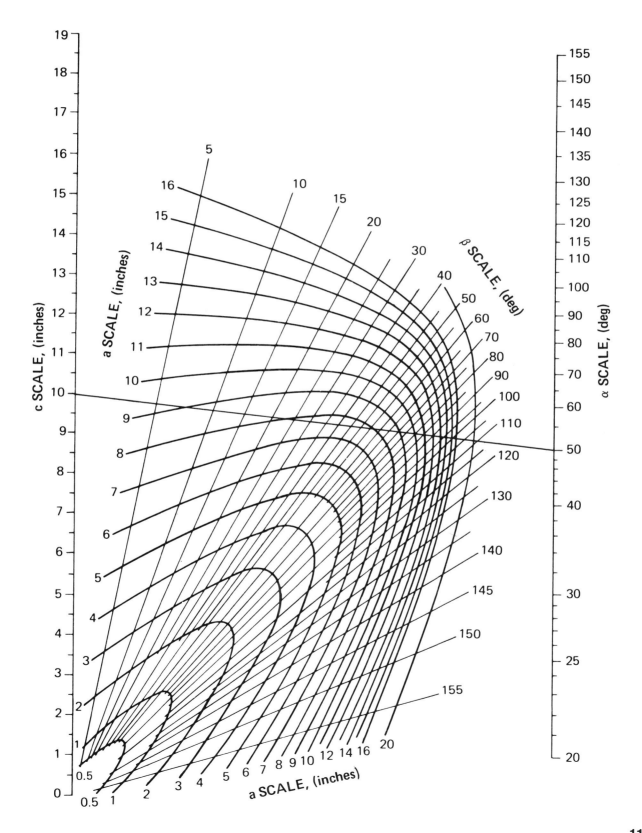

a SCALE, (inches)

c SCALE, (inches)

β SCALE, (deg)

α SCALE, (deg)

a SCALE, (inches)

111

Nomogram for law of cosines

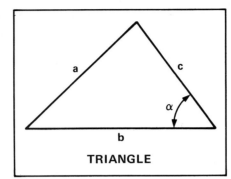

TRIANGLE

Introduction.

Frequently, triangles are solved by making use of the law of cosines:

$$a^2 = b^2 + c^2 - 2bc \cos \alpha \qquad (1)$$

The following nomogram provides a quick solution to Eq. 1.

Example 1:

Given a triangle that has $b = 6$ inches, $c = 2$ inches and $\alpha = 80$ deg, determine the length of side a of the triangle.

Solution 1:

Locate the intersection point of $c = 2$ and $b = 6$ on the upper c scale. Construct a line through this intersection point intersecting $\alpha = 80$ deg and the reference line. From this intersection with the reference line, construct a line to $c = 2$ on the lower c scale and where this line intersects the a scale, read the answer of $a = 6$ inches.

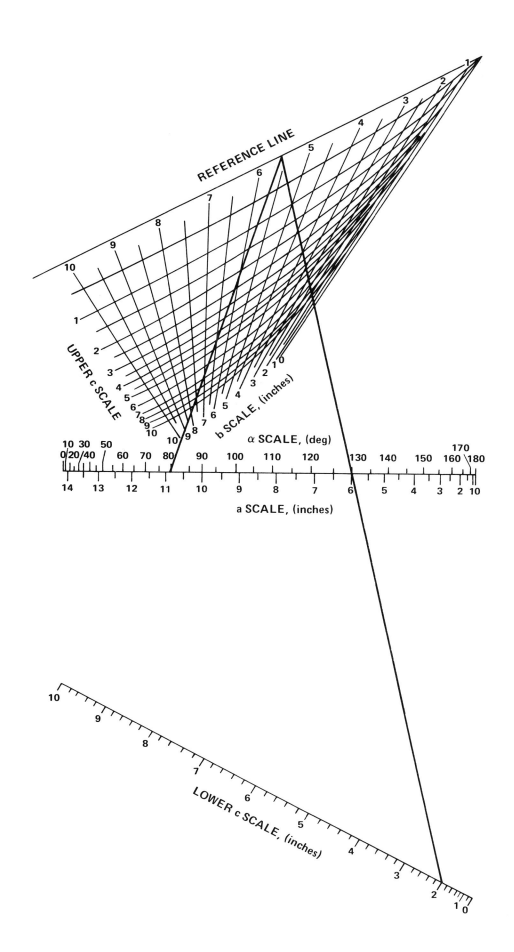

REFERENCE LINE

UPPER c SCALE

b SCALE, (inches)

α SCALE, (deg)

a SCALE, (inches)

LOWER c SCALE, (inches)

Decimal degree equivalents

Introduction.

Frequently in engineering calculations it is necessary to obtain a decimal degree equivalent of an angle. The following table expedites that process.

Example 1:

Given an angle of 54°45'34", determine its decimal degree equivalent.

MIN-UTES	SECONDS															
	0	1	2	3	4	5	6	7	8	9	10	11	12	13	14	15
0	0.0000	0.0003	0.0006	0.0008	0.0011	0.0014	0.0017	0.0019	0.0022	0.0025	0.0028	0.0031	0.0033	0.0036	0.0039	0.0042
1	.0167	.0169	.0172	.0175	.0178	.0181	.0183	.0186	.0189	.0192	.0194	.0197	.0200	.0203	.0206	.0208
2	.0333	.0336	.0339	.0342	.0344	.0347	.0350	.0353	.0356	.0358	.0361	.0364	.0367	.0369	.0372	.0375
3	.0500	.0503	.0506	.0508	.0511	.0514	.0517	.0519	.0522	.0525	.0528	.0531	.0533	.0536	.0539	.0542
4	.0667	.0669	.0672	.0675	.0678	.0681	.0683	.0686	.0689	.0692	.0694	.0697	.0700	.0703	.0706	.0708
5	.0833	.0836	.0839	.0842	.0844	.0847	.0850	.0853	.0856	.0858	.0861	.0864	.0867	.0869	.0872	.0875
6	.1000	.1003	.1006	.1008	.1011	.1014	.1017	.1019	.1022	.1025	.1028	.1031	.1033	.1036	.1039	.1042
7	.1167	.1169	.1172	.1175	.1178	.1181	.1183	.1186	.1189	.1192	.1194	.1197	.1200	.1203	.1206	.1208
8	.1333	.1336	.1339	.1342	.1344	.1347	.1350	.1353	.1356	.1358	.1361	.1364	.1367	.1369	.1372	.1375
9	.1500	.1503	.1506	.1508	.1511	.1514	.1517	.1519	.1522	.1525	.1528	.1531	.1533	.1536	.1539	.1542
10	.1667	.1669	.1672	.1675	.1678	.1681	.1683	.1686	.1689	.1692	.1694	.1697	.1700	.1703	.1706	.1708
11	.1833	.1836	.1839	.1842	.1844	.1847	.1850	.1853	.1856	.1858	.1861	.1864	.1867	.1869	.1872	.1875
12	.2000	.2003	.2006	.2008	.2011	.2014	.2017	.2019	.2022	.2025	.2028	.2031	.2033	.2036	.2039	.2042
13	.2167	.2169	.2172	.2175	.2178	.2181	.2183	.2186	.2189	.2192	.2194	.2197	.2200	.2203	.2206	.2208
14	.2333	.2336	.2339	.2342	.2344	.2347	.2350	.2353	.2356	.2358	.2361	.2364	.2367	.2369	.2372	.2375
15	.2500	.2503	.2506	.2508	.2511	.2514	.2517	.2519	.2522	.2525	.2528	.2531	.2533	.2536	.2539	.2542
16	.2667	.2669	.2672	.2675	.2678	.2681	.2683	.2686	.2689	.2692	.2694	.2697	.2700	.2703	.2706	.2708
17	.2833	.2836	.2839	.2842	.2844	.2847	.2850	.2853	.2856	.2858	.2861	.2864	.2867	.2869	.2872	.2875
18	.3000	.3003	.3006	.3008	.3011	.3014	.3017	.3019	.3022	.3025	.3028	.3031	.3033	.3036	.3039	.3042
19	.3167	.3169	.3172	.3175	.3178	.3181	.3183	.3186	.3189	.3192	.3194	.3197	.3200	.3203	.3206	.3208
20	.3333	.3336	.3339	.3342	.3344	.3347	.3350	.3353	.3356	.3358	.3361	.3364	.3367	.3369	.3372	.3375
21	.3500	.3503	.3506	.3508	.3511	.3514	.3517	.3519	.3522	.3525	.3528	.3531	.3533	.3536	.3539	.3542
22	.3667	.3669	.3672	.3675	.3678	.3681	.3683	.3686	.3689	.3692	.3694	.3697	.3700	.3703	.3706	.3708
23	.3833	.3836	.3839	.3842	.3844	.3847	.3850	.3853	.3856	.3858	.3861	.3864	.3867	.3869	.3872	.3875
24	.4000	.4003	.4006	.4008	.4011	.4014	.4017	.4019	.4022	.4025	.4028	.4031	.4033	.4036	.4039	.4042
25	.4167	.4169	.4172	.4175	.4178	.4181	.4183	.4186	.4189	.4192	.4194	.4197	.4200	.4203	.4206	.4208
26	.4333	.4336	.4339	.4342	.4344	.4347	.4350	.4353	.4356	.4358	.4361	.4364	.4367	.4369	.4372	.4375
27	.4500	.4503	.4506	.4508	.4511	.4514	.4517	.4519	.4522	.4525	.4528	.4531	.4533	.4536	.4539	.4542
28	.4667	.4669	.4672	.4675	.4678	.4681	.4683	.4686	.4689	.4692	.4694	.4697	.4700	.4703	.4706	.4708
29	.4833	.4836	.4839	.4842	.4844	.4847	.4850	.4853	.4856	.4858	.4861	.4864	.4867	.4869	.4872	.4875
30	.5000	.5003	.5006	.5008	.5011	.5014	.5017	.5019	.5022	.5025	.5028	.5031	.5033	.5036	.5039	.5042
31	.5167	.5169	.5172	.5175	.5178	.5181	.5183	.5186	.5189	.5192	.5194	.5197	.5200	.5203	.5206	.5208
32	.5333	.5336	.5339	.5342	.5344	.5347	.5350	.5353	.5356	.5358	.5361	.5364	.5367	.5369	.5372	.5375
33	.5500	.5503	.5506	.5508	.5511	.5514	.5517	.5519	.5522	.5525	.5528	.5531	.5533	.5536	.5539	.5542
34	.5667	.5669	.5672	.5675	.5678	.5681	.5683	.5686	.5689	.5692	.5694	.5697	.5700	.5703	.5706	.5708
35	.5833	.5836	.5839	.5842	.5844	.5847	.5850	.5853	.5856	.5858	.5861	.5864	.5867	.5869	.5872	.5875
36	.6000	.6003	.6006	.6008	.6011	.6014	.6017	.6019	.6022	.6025	.6028	.6031	.6033	.6036	.6039	.6042
37	.6167	.6169	.6172	.6175	.6178	.6181	.6183	.6186	.6189	.6192	.6194	.6197	.6200	.6203	.6206	.6208
38	.6333	.6336	.6339	.6342	.6344	.6347	.6350	.6353	.6356	.6358	.6361	.6364	.6367	.6369	.6372	.6375
39	.6500	.6503	.6506	.6508	.6511	.6514	.6517	.6519	.6522	.6525	.6528	.6531	.6533	.6536	.6539	.6542
40	.6667	.6669	.6672	.6675	.6678	.6681	.6683	.6686	.6689	.6692	.6694	.6697	.6700	.6703	.6706	.6708
41	.6833	.6836	.6839	.6842	.6844	.6847	.6850	.6853	.6856	.6858	.6861	.6864	.6867	.6869	.6872	.6875
42	.7000	.7003	.7006	.7008	.7011	.7014	.7017	.7019	.7022	.7025	.7028	.7031	.7033	.7036	.7039	.7042
43	.7167	.7169	.7172	.7175	.7178	.7181	.7183	.7186	.7189	.7192	.7194	.7197	.7200	.7203	.7206	.7208
44	.7333	.7336	.7339	.7342	.7344	.7347	.7350	.7353	.7356	.7358	.7361	.7364	.7367	.7369	.7372	.7375
45	.7500	.7503	.7506	.7508	.7511	.7514	.7517	.7519	.7522	.7525	.7528	.7531	.7533	.7536	.7539	.7542
46	.7667	.7669	.7672	.7675	.7678	.7681	.7683	.7686	.7689	.7692	.7694	.7697	.7700	.7703	.7706	.7708
47	.7833	.7836	.7839	.7842	.7844	.7847	.7850	.7853	.7856	.7858	.7861	.7864	.7867	.7869	.7872	.7875
48	.8000	.8003	.8006	.8008	.8011	.8014	.8017	.8019	.8022	.8025	.8028	.8031	.8033	.8036	.8039	.8042
49	.8167	.8169	.8172	.8175	.8178	.8181	.8183	.8186	.8189	.8192	.8194	.8197	.8200	.8203	.8206	.8208
50	.8333	.8336	.8339	.8342	.8344	.8347	.8350	.8353	.8356	.8358	.8361	.8364	.8367	.8369	.8372	.8375
51	.8500	.8503	.8506	.8508	.8511	.8514	.8517	.8519	.8522	.8525	.8528	.8531	.8533	.8536	.8539	.8542
52	.8667	.8669	.8672	.8675	.8678	.8681	.8683	.8686	.8689	.8692	.8694	.8697	.8700	.8703	.8706	.8708
53	.8833	.8836	.8839	.8842	.8844	.8847	.8850	.8853	.8856	.8858	.8861	.8864	.8867	.8869	.8872	.8875
54	.9000	.9003	.9006	.9008	.9011	.9014	.9017	.9019	.9022	.9025	.9028	.9031	.9033	.9036	.9039	.9042
55	.9167	.9169	.9172	.9175	.9178	.9181	.9183	.9186	.9189	.9192	.9194	.9197	.9200	.9203	.9206	.9208
56	.9333	.9336	.9339	.9342	.9344	.9347	.9350	.9353	.9356	.9358	.9361	.9364	.9367	.9369	.9372	.9375
57	.9500	.9503	.9506	.9508	.9511	.9514	.9517	.9519	.9522	.9525	.9528	.9531	.9533	.9536	.9539	.9542
58	.9667	.9669	.9672	.9675	.9678	.9681	.9683	.9686	.9689	.9692	.9694	.9697	.9700	.9703	.9706	.9708
59	.9833	.9836	9839	.9842	.9844	.9847	.9850	.9853	.9856	.9858	.9861	.9864	.9867	.9869	.9872	.9875

Solution 1:

Going to the table we locate 45 in the minute column. We proceed in a horizontal direction until we intersect the 34 second column. At this intersection point we read the answer of 0.7594. Thus the complete answer is:

$$54°45'34'' = 54.7594°$$

SECONDS

16	17	18	19	20	21	22	23	24	25	26	27	28	29	30	MIN-UTES
0.0044	0.0047	0.0050	0.0053	0.0056	0.0058	0.0061	0.0064	0.0067	0.0069	0.0072	0.0075	0.0078	0.0081	0.0083	0
.0211	.0214	.0217	.0219	.0222	.0225	.0228	.0231	.0233	.0236	.0239	.0242	.0244	.0247	.0250	1
.0378	.0381	.0383	.0386	.0389	.0392	.0394	.0397	.0400	.0403	.0406	.0408	.0411	.0414	.0417	2
.0544	.0547	.0550	.0553	.0556	.0558	.0561	.0564	.0567	.0569	.0572	.0575	.0578	.0581	.0583	3
.0711	.0714	.0717	.0719	.0722	.0725	.0728	.0731	.0733	.0736	.0739	.0742	.0744	.0747	.0750	4
.0878	.0881	.0883	.0886	.0889	.0892	.0894	.0897	.0900	.0903	.0906	.0908	.0911	.0914	.0917	5
.1044	.1047	.1050	.1053	.1056	.1058	.1061	.1064	.1067	.1069	.1072	.1075	.1078	.1081	.1083	6
.1211	.1214	.1217	.1219	.1222	.1225	.1228	.1231	.1233	.1236	.1239	.1242	.1244	.1247	.1250	7
.1378	.1381	.1383	.1386	.1389	.1392	.1394	.1397	.1400	.1403	.1406	.1408	.1411	.1414	.1417	8
.1544	.1547	.1550	.1553	.1556	.1558	.1561	.1564	.1567	.1569	.1572	.1575	.1578	.1581	.1583	9
.1711	.1714	.1717	.1719	.1722	.1725	.1728	.1731	.1733	.1736	.1739	.1742	.1744	.1747	.1750	10
.1878	.1881	.1883	.1886	.1889	.1892	.1894	.1897	.1900	.1903	.1906	.1908	.1911	.1914	.1917	11
.2044	.2047	.2050	.2053	.2056	.2058	.2061	.2064	.2067	.2069	.2072	.2075	.2078	.2081	.2083	12
.2211	.2214	.2217	.2219	.2222	.2225	.2228	.2231	.2233	.2236	.2239	.2242	.2244	.2247	.2250	13
.2378	.2381	.2383	.2386	.2389	.2392	.2394	.2397	.2400	.2403	.2406	.2408	.2411	.2414	.2417	14
.2544	.2547	.2550	.2553	.2556	.2558	.2561	.2564	.2567	.2569	.2572	.2575	.2578	.2581	.2583	15
.2711	.2714	.2717	.2719	.2722	.2725	.2728	.2731	.2733	.2736	.2739	.2742	.2744	.2747	.2750	16
.2878	.2881	.2883	.2886	.2889	.2892	.2894	.2897	.2900	.2903	.2906	.2908	.2911	.2914	.2917	17
.3044	.3047	.3050	.3053	.3056	.3058	.3061	.3064	.3067	.3069	.3072	.3075	.3078	.3081	.3083	18
.3211	.3214	.3217	.3219	.3222	.3225	.3228	.3231	.3233	.3236	.3239	.3242	.3244	.3247	.3250	19
.3378	.3381	.3383	.3386	.3389	.3392	.3394	.3397	.3400	.3403	.3406	.3408	.3411	.3414	.3417	20
.3544	.3547	.3550	.3553	.3556	.3558	.3561	.3564	.3567	.3569	.3572	.3575	.3578	.3581	.3583	21
.3711	.3714	.3717	.3719	.3722	.3725	.3728	.3731	.3733	.3736	.3739	.3742	.3744	.3747	.3750	22
.3878	.3881	.3883	.3886	.3889	.3892	.3894	.3897	.3900	.3903	.3906	.3908	.3911	.3914	.3917	23
.4044	.4047	.4050	.4053	.4056	.4058	.4061	.4064	.4067	.4069	.4072	.4075	.4078	.4081	.4083	24
.4211	.4214	.4217	.4219	.4222	.4225	.4228	.4231	.4233	.4236	.4239	.4242	.4244	.4247	.4250	25
.4378	.4381	.4383	.4386	.4389	.4392	.4394	.4397	.4400	.4403	.4406	.4408	.4411	.4414	.4417	26
.4544	.4547	.4550	.4553	.4556	.4558	.4561	.4564	.4567	.4569	.4572	.4575	.4578	.4581	.4583	27
.4711	.4714	.4717	.4719	.4722	.4725	.4728	.4731	.4733	.4736	.4739	.4742	.4744	.4747	.4750	28
.4878	.4881	.4883	.4886	.4889	.4892	.4894	.4897	.4900	.4903	.4906	.4908	.4911	.4914	.4917	29
.5044	.5047	.5050	.5053	.5056	.5058	.5061	.5064	.5067	.5069	.5072	.5075	.5078	.5081	.5083	30
.5211	.5214	.5217	.5219	.5222	.5225	.5228	.5231	.5233	.5236	.5239	.5242	.5244	.5247	.5250	31
.5378	.5381	.5383	.5386	.5389	.5392	.5394	.5397	.5400	.5403	.5406	.5408	.5411	.5414	.5417	32
.5544	.5547	.5550	.5553	.5556	.5558	.5561	.5564	.5567	.5569	.5572	.5575	.5578	.5581	.5583	33
.5711	.5714	.5717	.5719	.5722	.5725	.5728	.5731	.5733	.5736	.5739	.5742	.5744	.5747	.5750	34
.5878	.5881	.5883	.5886	.5889	.5892	.5894	.5897	.5900	.5903	.5906	.5908	.5911	.5914	.5917	35
.6044	.6047	.6050	.6053	.6056	.6058	.6061	.6064	.6067	.6069	.6072	.6075	.6078	.6081	.6083	36
.6211	.6214	.6217	.6219	.6222	.6225	.6228	.6231	.6233	.6236	.6239	.6242	.6244	.6247	.6250	37
.6378	.6381	.6383	.6386	.6389	.6392	.6394	.6397	.6400	.6403	.6406	.6408	.6411	.6414	.6417	38
.6544	.6547	.6550	.6553	.6556	.6558	.6561	.6564	.6567	.6569	.6572	.6575	.6578	.6581	.6583	39
.6711	.6714	.6717	.6719	.6722	.6725	.6728	.6731	.6733	.6736	.6739	.6742	.6744	.6747	.6750	40
.6878	.6881	.6883	.6886	.6889	.6892	.6894	.6897	.6900	.6903	.6906	.6908	.6911	.6914	.6917	41
.7044	.7047	.7050	.7053	.7056	.7058	.7061	.7064	.7067	.7069	.7072	.7075	.7078	.7081	.7083	42
.7211	.7214	.7217	.7219	.7222	.7225	.7228	.7231	.7233	.7236	.7239	.7242	.7244	.7247	.7250	43
.7378	.7381	.7383	.7386	.7389	.7392	.7394	.7397	.7400	.7403	.7406	.7408	.7411	.7414	.7417	44
.7544	.7547	.7550	.7553	.7556	.7558	.7561	.7564	.7567	.7569	.7572	.7575	.7578	.7581	.7583	45
.7711	.7714	.7717	.7719	.7722	.7725	.7728	.7731	.7733	.7736	.7739	.7742	.7744	.7747	.7750	46
.7878	.7881	.7883	.7886	.7889	.7892	.7894	.7897	.7900	.7903	.7906	.7908	.7911	.7914	.7917	47
.8044	.8047	.8050	.8053	.8056	.8058	.8061	.8064	.8067	.8069	.8072	.8075	.8078	.8081	.8083	48
.8211	.8214	.8217	.8219	.8222	.8225	.8228	.8231	.8233	.8236	.8239	.8242	.8244	.8247	.8250	49
.8378	.8381	.8383	.8386	.8389	.8392	.8394	.8397	.8400	.8403	.8406	.8408	.8411	.8414	.8417	50
.8544	.8547	.8550	.8553	.8556	.8558	.8561	.8564	.8567	.8569	.8572	.8575	.8578	.8581	.8583	51
.8711	.8714	.8717	.8719	.8722	.8725	.8728	.8731	.8733	.8736	.8739	.8742	.8744	.8747	.8750	52
.8878	.8881	.8883	.8886	.8889	.8892	.8894	.8897	.8900	.8903	.8906	.8908	.8911	.8914	.8917	53
.9044	.9047	.9050	.9053	.9056	.9058	.9061	.9064	.9067	.9069	.9072	.9075	.9078	.9081	.9083	54
.9211	.9214	.9217	.9219	.9222	.9225	.9228	.9231	.9233	.9236	.9239	.9242	.9244	.9247	.9250	55
.9378	.9381	.9383	.9386	.9389	.9392	.9394	.9397	.9400	.9403	.9406	.9408	.9411	.9414	.9417	56
.9544	.9547	.9550	.9553	.9556	.9558	.9561	.9564	.9567	.9569	.9572	.9575	.9578	.9581	.9583	57
.9711	.9714	.9717	.9719	.9722	.9725	.9728	.9731	.9733	.9736	.9739	.9742	.9744	.9747	.9750	58
.9878	.9881	.9883	.9886	.9889	.9892	.9894	.9897	.9900	.9903	.9906	.9908	.9911	.9914	.9917	59

MIN-UTES	31	32	33	34	35	36	37	38	39	40	41	42	43	44	45
0	0.0086	0.0089	0.0092	0.0094	0.0097	0.0100	0.0103	0.0106	0.0108	0.0111	0.0114	0.0117	0.0119	0.0122	0.0125
1	.0253	.0256	.0258	.0261	.0264	.0267	.0269	.0272	.0275	.0278	0.281	.0283	.0286	.0289	.0292
2	.0419	.0422	.0425	.0428	.0431	.0433	.0436	.0439	.0442	.0444	.0447	.0450	.0453	.0456	.0458
3	.0586	.0589	.0592	.0594	.0597	.0600	.0603	.0606	.0608	.0611	.0614	.0617	.0619	.0622	.0625
4	.0753	.0756	.0758	.0761	.0764	.0767	.0769	.0772	.0775	.0778	.0781	.0783	.0786	.0789	.0792
5	.0919	.0922	.0925	.0928	.0931	.0933	.0936	.0939	.0942	.0944	.0947	.0950	.0953	.0956	.0958
6	.1086	.1089	.1092	.1094	.1097	.1100	.1103	.1106	.1108	.1111	.1114	.1117	.1119	.1122	.1125
7	.1253	.1256	.1258	.1261	.1264	.1267	.1269	.1272	.1275	.1278	.1281	.1283	.1286	.1289	.1292
8	.1419	.1422	.1425	.1428	.1431	.1433	.1436	.1439	.1442	.1444	.1447	.1450	.1453	.1456	.1458
9	.1586	.1589	.1592	.1594	.1597	.1600	.1603	.1606	.1608	.1611	.1614	.1617	.1619	.1622	.1625
10	.1753	.1756	.1758	.1761	.1764	.1767	.1769	.1772	.1775	.1778	.1781	.1783	.1786	.1789	.1792
11	.1919	.1922	.1925	.1928	.1931	.1933	.1936	.1939	.1942	.1944	.1947	.1950	.1953	.1956	.1958
12	.2086	.2089	.2092	.2094	.2097	.2100	.2103	.2106	.2108	.2111	.2114	.2117	.2119	.2122	.2125
13	.2253	.2256	.2258	.2261	.2264	.2267	.2269	.2272	.2275	.2278	.2281	.2283	.2286	.2289	.2292
14	.2419	.2422	.2425	.2428	.2431	.2433	.2436	.2439	.2442	.2444	.2447	.2450	.2453	.2456	.2458
15	.2586	.2589	.2592	.2594	.2597	.2600	.2603	.2606	.2608	.2611	.2614	.2617	.2619	.2622	.2625
16	.2753	.2756	.2758	.2761	.2764	.2767	.2769	.2772	.2775	.2778	.2781	.2783	.2786	.2789	.2792
17	.2919	.2922	.2925	.2928	.2931	.2933	.2936	.2939	.2942	.2944	.2947	.2950	.2953	.2956	.2958
18	.3086	.3089	.3092	.3094	.3097	.3100	.3103	.3106	.3108	.3111	.3114	.3117	.3119	.3122	.3125
19	.3253	.3256	.3258	.3261	.3264	.3267	.3269	.3272	.3275	.3278	.3281	.3283	.3286	.3289	.3292
20	.3419	.3422	.3425	.3428	.3431	.3433	.3436	.3439	.3442	.3444	.3447	.3450	.3453	.3456	.3458
21	.3586	.3589	.3592	.3594	.3597	.3600	.3603	.3606	.3608	.3611	.3614	.3617	.3619	.3622	.3625
22	.3753	.3756	.3758	.3761	.3764	.3767	.3769	.3772	.3775	.3778	.3781	.3783	.3786	.3789	.3792
23	.3919	.3922	.3925	.3928	.3931	.3933	.3936	.3939	.3942	.3944	.3947	.3950	.3953	.3956	.3958
24	.4086	.4089	.4092	.4094	.4097	.4100	.4103	.4106	.4108	.4111	.4114	.4117	.4119	.4122	.4125
25	.4253	.4256	.4258	.4261	.4264	.4267	.4269	.4272	.4275	.4278	.4281	.4283	.4286	.4289	.4292
26	.4419	.4422	.4425	.4428	.4431	.4433	.4436	.4439	.4442	.4444	.4447	.4450	.4453	.4456	.4458
27	.4586	.4589	.4592	.4594	.4597	.4600	.4603	.4606	.4608	.4611	.4614	.4617	.4619	.4622	.4625
28	.4753	.4756	.4758	.4761	.4764	.4767	.4769	.4772	.4775	.4778	.4781	.4783	.4786	.4789	.4792
29	.4919	.4922	.4925	.4928	.4931	.4933	.4936	.4939	.4942	.4944	.4947	.4950	.4953	.4956	.4958
30	.5086	.5089	.5092	.5094	.5097	.5100	.5103	.5106	.5108	.5111	.5114	.5117	.5119	.5122	.5125
31	.5253	.5256	.5258	.5261	.5264	.5267	.5269	.5272	.5275	.5278	.5281	.5283	.5286	.5289	.5292
32	.5419	.5422	.5425	.5428	.5431	.5433	.5436	.5439	.5442	.5444	.5447	.5450	.5453	.5456	.5458
33	.5586	.5589	.5592	.5594	.5597	.5600	.5603	.5606	.5608	.5611	.5614	.5617	.5619	.5622	.5625
34	.5753	.5756	.5758	.5761	.5764	.5767	.5769	.5772	.5775	.5778	.5781	.5783	.5786	.5789	.5792
35	.5919	.5922	.5925	.5928	.5931	.5933	.5936	.5939	.5942	.5944	.5947	.5950	.5953	.5956	.5958
36	.6086	.6089	.6092	.6094	.6097	.6100	.6103	.6106	.6108	.6111	.6114	.6117	.6119	.6122	.6125
37	.6253	.6256	.6258	.6261	.6264	.6267	.6269	.6272	.6275	.6278	.6281	.6283	.6286	.6289	.6292
38	.6419	.6422	.6425	.6428	.6431	.6433	.6436	.6439	.6442	.6444	.6447	.6450	.6453	.6456	.6458
39	.6586	.6589	.6592	.6594	.6597	.6600	.6603	.6606	.6608	.6611	.6614	.6617	.6619	.6622	.6625
40	.6753	.6756	.6758	.6761	.6764	.6767	.6769	.6772	.6775	.6778	.6781	.6783	.6786	.6789	.6792
41	.6919	.6922	.6925	.6928	.6931	.6933	.6936	.6939	.6942	.6944	.6947	.6950	.6953	.6956	.6958
42	.7086	.7089	.7092	.7094	.7097	.7100	.7103	.7106	.7108	.7111	.7114	.7117	.7119	.7122	.7125
43	.7253	.7256	.7258	.7261	.7264	.7267	.7269	.7272	.7275	.7278	.7281	.7283	.7286	.7289	.7292
44	.7419	.7422	.7425	.7428	.7431	.7433	.7436	.7439	.7442	.7444	.7447	.7450	.7453	.7456	.7458
45	.7586	.7589	.7592	.7594	.7597	.7600	.7603	.7606	.7608	.7611	.7614	.7617	.7619	.7622	.7625
46	.7753	.7756	.7758	.7761	.7764	.7767	.7769	.7772	.7775	.7778	.7781	.7783	.7786	.7789	.7792
47	.7919	.7922	.7925	.7928	.7931	.7933	.7936	.7939	.7942	.7944	.7947	.7950	.7953	.7956	.7958
48	.8086	.8089	.8092	.8094	.8097	.8100	.8103	.8106	.8108	.8111	.8114	.8117	.8119	.8122	.8125
49	.8253	.8256	.8258	.8261	.8264	.8267	.8269	.8272	.8275	.8278	.8281	.8283	.8286	.8289	.8292
50	.8419	.8422	.8425	.8428	.8431	.8433	.8436	.8439	.8442	.8444	.8447	.8450	.8453	.8456	.8458
51	.8586	.8589	.8592	.8594	.8597	.8600	.8603	.8606	.8608	.8611	.8614	.8617	.8619	.8622	.8625
52	.8753	.8756	.8758	.8761	.8764	.8767	.8769	.8772	.8775	.8778	.8781	.8783	.8786	.8789	.8792
53	.8919	.8922	.8925	.8928	.8931	.8933	.8936	.8939	.8942	.8944	.8947	.8950	.8953	.8956	.8958
54	.9086	.9089	.9092	.9094	.9097	.9100	.9103	.9106	.9108	.9111	.9114	.9117	.9119	.9122	.9125
55	.9253	.9256	.9258	.9261	.9264	.9267	.9269	.9272	.9275	.9278	.9281	.9283	.9286	.9289	.9292
56	.9419	.9422	.9425	.9428	.9431	.9433	.9436	.9439	.9442	.9444	.9447	.9450	.9453	.9456	.9458
57	.9586	.9589	.9592	.9594	.9597	.9600	.9603	.9606	.9608	.9611	.9614	.9617	.9619	.9622	.9625
58	.9753	.9756	.9758	.9761	.9764	.9767	.9769	.9772	.9775	.9778	.9781	.9783	.9786	.9789	.9792
59	.9919	.9922	.9925	.9928	.9931	.9933	.9936	.9939	.9942	.9944	.9947	.9950	.9953	.9956	.9958

46	47	48	49	50	51	52	53	54	55	56	57	58	59	MIN-UTES
0.0128	0.0131	0.0133	0.0136	0.0139	0.0142	0.0144	0.0147	0.0150	0.0153	0.0156	0.0158	0.0161	0.0164	0
.0294	.0297	.0300	.0303	.0306	.0308	.0311	.0314	.0317	.0319	.0322	0.0325	0.0328	.0331	1
.0461	.0464	.0467	.0469	.0472	.0475	.0478	.0481	.0483	.0486	.0489	.0492	.0494	.0497	2
.0628	.0631	.0633	.0636	.0639	.0642	.0644	.0647	.0650	.0653	.0656	.0658	.0661	.0664	3
.0794	.0797	.0800	.0803	.0806	.0808	.0811	.0814	.0817	.0819	.0822	.0825	.0828	.0831	4
.0961	.0964	.0967	.0969	.0972	.0975	.0978	.0981	.0983	.0986	.0989	.0992	.0994	.0997	5
.1128	.1131	.1133	.1136	.1139	.1142	.1144	.1147	.1150	.1153	.1156	.1158	.1161	.1164	6
.1294	.1297	.1300	.1303	.1306	.1308	.1311	.1314	.1317	.1319	.1322	.1325	.1328	.1331	7
.1461	.1464	.1467	.1469	.1472	.1475	.1478	.1481	.1483	.1486	.1489	.1492	.1494	.1497	8
.1628	.1631	.1633	.1636	.1639	.1642	.1644	.1647	.1650	.1653	.1656	.1658	.1661	.1664	9
.1794	.1797	.1800	.1803	.1806	.1808	.1811	.1814	.1827	.1819	.1822	.1825	.1828	.1831	10
.1961	.1964	.1967	.1969	.1972	.1975	.1978	.1981	.1983	.1986	.1989	.1992	.1994	.1997	11
.2128	.2131	.2133	.2136	.2139	.2142	.2144	.2147	.2150	.2153	.2156	.2158	.2161	.2164	12
.2294	.2297	.2300	.2303	.2306	.2308	.2311	.2314	.2317	.2319	.2322	.2325	.2328	.2331	13
.2461	.2464	.2467	.2469	.2472	.2475	.2478	.2481	.2483	.2486	.2489	.2492	.2494	.2497	14
.2628	.2631	.2633	.2636	.2639	.2642	.2644	.2647	.2650	.2653	.2656	.2658	.2661	.2664	15
.2794	.2797	.2800	.2803	.2806	.2808	.2811	.2814	.2817	.2819	.2822	.2825	.2828	.2831	16
.2961	.2964	.2967	.2969	.2972	.2975	.2978	.2981	.2983	.2986	.2989	.2992	.2994	.2997	17
.3128	.3131	.3133	.3136	.3139	.3142	.3144	.3147	.3150	.3153	.3156	.3158	.3161	.3164	18
.3294	.3297	.3300	.3303	.3306	.3308	.3311	.3314	.3317	.3319	.3322	.3325	.3328	.3331	19
.3461	.3464	.3467	.3469	.3472	.3475	.3478	.3481	.3483	.3486	.3489	.3492	.3494	.3497	20
.3628	.3631	.3633	.3636	.3639	.3642	.3644	.3647	.3650	.3653	.3656	.3658	.3661	.3664	21
.3794	.3797	.3800	.3803	.3806	.3808	.3811	.3814	.3817	.3819	.3822	.3825	.3828	.3831	22
.3961	.3964	.3967	.3969	.3972	.3975	.3978	.3981	.3983	.3986	.3989	.3992	.3994	.3997	23
.4128	.4131	.4133	.4136	.4139	.4142	.4144	.4147	.4150	.4153	.4156	.4158	.4161	.4164	24
.4294	.4297	.4300	.4303	.4306	.4308	.4311	.4314	.4317	.4319	.4322	.4325	.4328	.4331	25
.4461	.4464	.4467	.4469	.4472	.4475	.4478	.4481	.4483	.4486	.4489	.4492	.4494	.4497	26
.4628	.4631	.4633	.4636	.4639	.4642	.4644	.4647	.4650	.4653	.4656	.4658	.4661	.4664	27
.4794	.4797	.4800	.4803	.4806	.4808	.4811	.4814	.4817	.4819	.4822	.4825	.4828	.4831	28
.4961	.4964	.4967	.4969	.4972	.4975	.4978	.4981	.4983	.4986	.4989	.4992	.4994	.4997	29
.5128	.5131	.5133	.5136	.5139	.5142	.5144	.5147	.5150	.5153	.5156	.5158	.5161	.5164	30
.5294	.5297	.5300	.5303	.5306	.5308	.5311	.5314	.5317	.5319	.5322	.5325	.5328	.5331	31
.5461	.5464	.5467	.5469	.5472	.5475	.5478	.5481	.5483	.5486	.5489	.5492	.5494	.5497	32
.5628	.5631	.5633	.5636	.5639	.5642	.5644	.5647	.5650	.5653	.5656	.5658	.5661	.5664	33
.5794	.5797	.5800	.5803	.5806	.5808	.5811	.5814	.5817	.5819	.5822	.5825	.5828	.5831	34
.5961	.5964	.5967	.5969	.5972	.5975	.5978	.5981	.5983	.5986	.5989	.5992	.5994	.5997	35
.6128	.6131	.6133	.6136	.6139	.6142	.6144	.6147	.6150	.6153	.6156	.6158	.6161	.6164	36
.6294	.6297	.6300	.6303	.6306	.6308	.6311	.6314	.6317	.6319	.6322	.6325	.6328	.6331	37
.6461	.6464	.6467	.6469	.6472	.6475	.6478	.6481	.6483	.6486	.6489	.6492	.6494	.6497	38
.6628	.6631	.6633	.6636	.6639	.6642	.6644	.6647	.6650	.6653	.6656	.6658	.6661	.6664	39
.6794	.6797	.6800	.6803	.6806	.6808	.6811	.6814	.6817	.6819	.6822	.6825	.6828	.6831	40
.6961	.6964	.6967	.6969	.6972	.6975	.6978	.6981	.6983	.6986	.6989	.6992	.6994	.6997	41
.7128	.7131	.7133	.7136	.7139	.7142	.7144	.7147	.7150	.7153	.7156	.7158	.7161	.7164	42
.7294	.7297	.7300	.7303	.7306	.7308	.7311	.7314	.7317	.7319	.7322	.7325	.7328	.7331	43
.7461	.7464	.7467	.7469	.7472	.7475	.7478	.7481	.7483	.7486	.7489	.7492	.7494	.7497	44
.7628	.7631	.7633	.7636	.7639	.7642	.7644	.7647	.7650	.7653	.7656	.7658	.7661	.7664	45
.7794	.7797	.7800	.7803	.7806	.7808	.7811	.7814	.7817	.7819	.7822	.7825	.7828	.7831	46
.7961	.7964	.7967	.7969	.7972	.7975	.7978	.7981	.7983	.7986	.7989	.7992	.7994	.7997	47
.8128	.8131	.8133	.8136	.8139	.8142	.8144	.8147	.8150	.8153	.8156	.8158	.8161	.8164	48
.8294	.8297	.8300	.8303	.8306	.8308	.8311	.8314	.8317	.8319	.8322	.8325	.8328	.8331	49
.8461	.8464	.8467	.8469	.8472	.8475	.8478	.8481	.8483	.8486	.8489	.8492	.8494	.8497	50
.8628	.8631	.8633	.8636	.8639	.8642	.8644	.8647	.8650	.8653	.8656	.8658	.8661	.8664	51
.8794	.8797	.8800	.8803	.8806	.8808	.8811	.8814	.8817	.8819	.8822	.8825	.8828	.8831	52
.8961	.8964	.8967	.8969	.8972	.8975	.8978	.8981	.8983	.8986	.8989	.8992	.8994	.8997	53
.9128	.9131	.9133	.9136	.9139	.9142	.9144	.9147	.9150	.9153	.9156	.9158	.9161	.9164	54
.9294	.9297	.9300	.9303	.9306	.9308	.9311	.9314	.9317	.9319	.9322	.9325	.9328	.9331	55
.9461	.9464	.9467	.9469	.9472	.9475	.9478	.9481	.9483	.9486	.9489	.9492	.9494	.9497	56
.9628	.9631	.9633	.9636	.9639	.9642	.9644	.9647	.9650	.9653	.9656	.9658	.9661	.9664	57
.9794	.9797	.9800	.9803	.9806	.9808	.9811	.9814	.9817	.9819	.9822	.9825	.9828	.9831	58
.9961	.9964	.9967	.9969	.9972	.9975	.9978	.9981	.9983	.9986	.9989	.9992	.9994	.9997	59

Degree-radian comparisons

Introduction.
Frequently in the solution of engineering problems, the relationship between degrees and radians is required. The following graph facilitates this task.

Example 1:
Determine the radian equivalent of 57.3 deg.
Solution 1:
Locate 57.3 deg on the deg scale and read the answer of 57.3 deg = 1.0 radian.

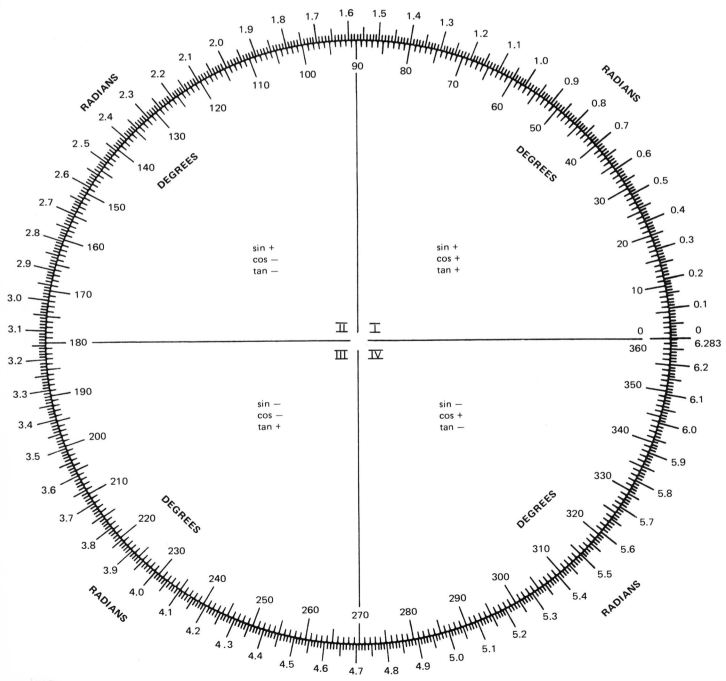

sin +
cos −
tan −

sin +
cos +
tan +

II | I

III | IV

sin −
cos −
tan +

sin −
cos +
tan −

Angular relationships

Introduction.
Frequently in engineering calculations it is necessary to determine the specific functions related to different angles. The following tables give many different angles and the specific functions related to these angles.

Example 1:
Given an angle of 60 deg, determine its Arc, Sin, Cos, Tan, Cot, Sec, Csc and Chord.

Solution 1:
Going to the tables and locating 60 deg, read the following answers:

$$\text{Angle (rad)} = \frac{1}{3}\pi$$

$$\text{Sin} = \frac{1}{2}\sqrt{3}$$

$$\text{Cos} = \frac{1}{2}$$

$$\text{Tan} = \sqrt{3}$$

$$\text{Cot} = \frac{1}{3}\sqrt{3}$$

$$\text{Sec} = 2$$

$$\text{Csc} = \frac{2}{3}\sqrt{3}$$

$$\text{Chord} = 1$$

Angle (deg)	Angle (rad)	Sin	Cos	Tan	Cot	Sec	Csc	Chord
0	0	0	+1	0	∞	+1	∞	0
30	$\frac{1}{6}\pi$	$1/2$	$\frac{1}{2}\sqrt{3}$	$\frac{1}{3}\sqrt{3}$	$\sqrt{3}$	$\frac{2}{3}\sqrt{3}$	2	$\sqrt{2-\sqrt{3}}$
45	$\frac{1}{4}\pi$	$\frac{1}{2}\sqrt{2}$	$\frac{1}{2}\sqrt{2}$	$+1$	$+1$	$\sqrt{2}$	$\sqrt{2}$	$\sqrt{2-\sqrt{2}}$
60	$\frac{1}{3}\pi$	$\frac{1}{2}\sqrt{3}$	$1/2$	$\sqrt{3}$	$\frac{1}{3}\sqrt{3}$	2	$\frac{2}{3}\sqrt{3}$	1
90	$\frac{1}{2}\pi$	$+1$	0	∞	0	∞	$+1$	$\sqrt{2}$
120	$\frac{2}{3}\pi$	$\frac{1}{2}\sqrt{3}$	$-1/2$	$-\sqrt{3}$	$-\frac{1}{3}\sqrt{3}$	-2	$\frac{2}{3}\sqrt{3}$	$\sqrt{3}$
135	$\frac{3}{4}\pi$	$\frac{1}{2}\sqrt{2}$	$-\frac{1}{2}\sqrt{2}$	-1	-1	$-\sqrt{2}$	$\sqrt{2}$	$\sqrt{2+\sqrt{2}}$
150	$\frac{5}{6}\pi$	$1/2$	$-\frac{1}{2}\sqrt{3}$	$-\frac{1}{3}\sqrt{3}$	$-\sqrt{3}$	$-\frac{2}{3}\sqrt{3}$	2	$\sqrt{2+\sqrt{3}}$
180	π	0	-1	0	∞	-1	∞	2
210	$\frac{7}{6}\pi$	$-1/2$	$-\frac{1}{2}\sqrt{3}$	$\frac{1}{3}\sqrt{3}$	$\sqrt{3}$	$-\frac{2}{3}\sqrt{3}$	-2	$\sqrt{2+\sqrt{3}}$
225	$\frac{5}{4}\pi$	$-\frac{1}{2}\sqrt{2}$	$-\frac{1}{2}\sqrt{2}$	$+1$	$+1$	$-\sqrt{2}$	$-\sqrt{2}$	$\sqrt{2+\sqrt{2}}$
240	$\frac{4}{3}\pi$	$-\frac{1}{2}\sqrt{3}$	$-1/2$	$\sqrt{3}$	$\frac{1}{3}\sqrt{3}$	-2	$-\frac{2}{3}\sqrt{3}$	$\sqrt{3}$
270	$\frac{3}{2}\pi$	-1	0	∞	0	∞	-1	$\sqrt{2}$
300	$\frac{5}{3}\pi$	$-\frac{1}{2}\sqrt{3}$	$1/2$	$-\sqrt{3}$	$-\frac{1}{3}\sqrt{3}$	2	$-\frac{2}{3}\sqrt{3}$	1
315	$\frac{7}{4}\pi$	$-\frac{1}{2}\sqrt{2}$	$\frac{1}{2}\sqrt{2}$	-1	-1	$\sqrt{2}$	$-\sqrt{2}$	$\sqrt{2-\sqrt{2}}$
330	$\frac{11}{6}\pi$	$-1/2$	$\frac{1}{2}\sqrt{3}$	$-\frac{1}{3}\sqrt{3}$	$-\sqrt{3}$	$\frac{2}{3}\sqrt{3}$	-2	$\sqrt{2-\sqrt{3}}$
360	2π	0	$+1$	0	∞	$+1$	∞	0

Read down (deg)	Sin	Cos	Tan	Cot	Sec	Csc	
$\frac{1}{12}\pi = 15$	$\dfrac{\sqrt{6}-\sqrt{2}}{4}$	$\dfrac{\sqrt{6}+\sqrt{2}}{4}$	$2-\sqrt{3}$	$2+\sqrt{3}$	$\sqrt{6}-\sqrt{2}$	$\sqrt{6}+\sqrt{2}$	$\frac{5}{12}\pi = 75$
$\frac{1}{10}\pi = 18$	$\dfrac{\sqrt{5}-1}{4}$	$\dfrac{\sqrt{10+2\sqrt{5}}}{4}$	$\dfrac{\sqrt{25-10\sqrt{5}}}{5}$	$\sqrt{5+2\sqrt{5}}$	$\dfrac{\sqrt{50-10\sqrt{5}}}{5}$	$\sqrt{5}+1$	$\frac{2}{5}\pi = 72$
$\frac{1}{6}\pi = 30$	$\dfrac{1}{2}$	$\dfrac{\sqrt{3}}{2}$	$\dfrac{\sqrt{3}}{3}$	$\sqrt{3}$	$\dfrac{2\sqrt{3}}{3}$	2	$\frac{1}{3}\pi = 60$
$\frac{1}{5}\pi = 36$	$\dfrac{\sqrt{10-2\sqrt{5}}}{4}$	$\dfrac{\sqrt{5}+1}{4}$	$\sqrt{5-2\sqrt{5}}$	$\dfrac{\sqrt{25+10\sqrt{5}}}{5}$	$\sqrt{5}-1$	$\dfrac{\sqrt{50+10\sqrt{5}}}{5}$	$\frac{3}{10}\pi = 54$
$\frac{1}{4}\pi = 45$	$\dfrac{\sqrt{2}}{2}$	$\dfrac{\sqrt{2}}{2}$	1	1	$\sqrt{2}$	$\sqrt{2}$	$\frac{1}{4}\pi = 45$
	Cos	Sin	Cot	Tan	Csc	Sec	Read up (deg)

Trigonometric short cut

Introduction.

The following trigonometric circle provides a fast, simple method of determining values of the sine, cosine and tangent of angles.

Example 1:

What is the sine of 45 deg?

Solution 1:

On the degree scale, locate 45 deg and follow this radial line until it intersects the sine scale. At this intersection point read the answer of sine 45 deg = 0.710.

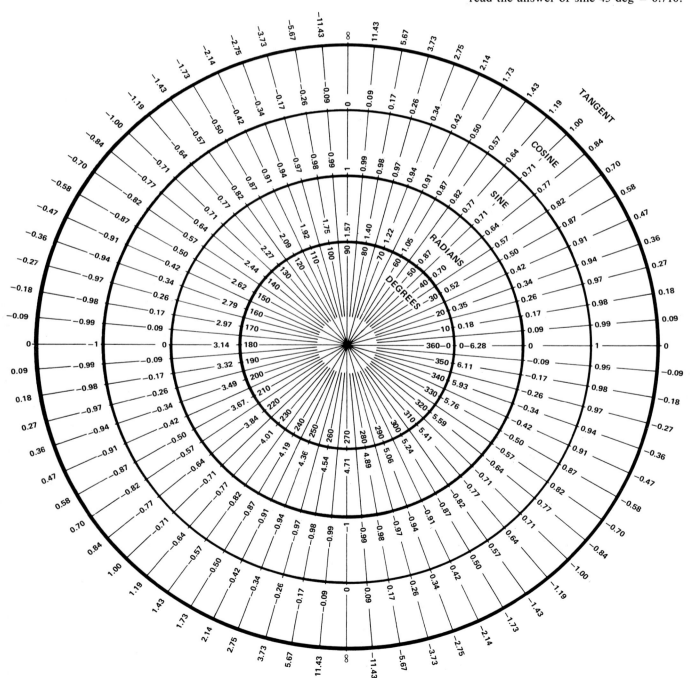

120

Short method for trigonometric solutions

Introduction.

The solution of many problems can be expedited by the manipulation of basic trigonometric relationships. The following set of triangles facilitates the expression of any function in terms of the others, as well as the determination of all the functions in terms of a given one.

Example 1:

Express the sine in terms of the other functions.

Solution 1:

From the set of triangles:

$\sin \theta = \sqrt{1 - \cos^2\theta}$

$\sin \theta = \dfrac{\tan \theta}{\sqrt{1 + \tan^2\theta}}$

$\sin \theta = \dfrac{1}{\sqrt{1 + \cot^2\theta}}$

$\sin \theta = \dfrac{\sqrt{\sec^2\theta - 1}}{\sec \theta}$

$\sin \theta = \dfrac{1}{\csc \theta}$

Example 2:

Express all of the functions in terms of the sine.

Solution 2:

From the set of triangles:

$\cos \theta = \sqrt{1 - \sin^2\theta}$

$\tan \theta = \dfrac{\sin \theta}{\sqrt{1 - \sin^2\theta}}$

$\cot \theta = \dfrac{\sqrt{1 - \sin^2\theta}}{\sin \theta}$

$\sec \theta = \dfrac{1}{\sqrt{1 - \sin^2\theta}}$

$\csc \theta = \dfrac{1}{\sin \theta}$

Example 3:

Given the equation, $3 \sin^2\theta + 5 \tan^2\theta = 2$, determine the angle θ.

Solution 3:

From the sine triangle, substituting the tangent in terms of the sine:

$3 \sin^2\theta + 5 \dfrac{\sin^2\theta}{1 - \sin^2\theta} = 2$

Solving this equation:

$\theta = \sin^{-1}\sqrt{\dfrac{5 \pm \sqrt{19}}{3}}$

Example 4:

Given $\sin \theta = 0.40$, determine $\tan \theta$.

Solution 4:

From the sine triangle, substituting the sine in terms of the tangent:

$\tan \theta = \dfrac{0.40}{\sqrt{1 - 0.40^2}}$

$\tan \theta = 0.440$

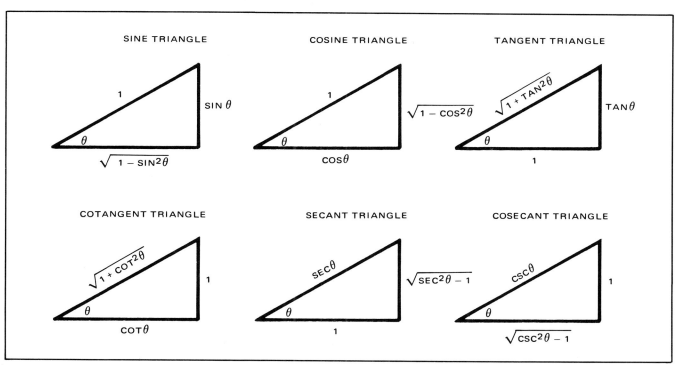

SINE TRIANGLE

COSINE TRIANGLE

TANGENT TRIANGLE

COTANGENT TRIANGLE

SECANT TRIANGLE

COSECANT TRIANGLE

Short method for hyperbolic solutions

Introduction.
The solution of many problems can be expedited by the manipulation of basic hyperbolic relationships. The following set of triangles facilitates the expression of any function in terms of the others, as well as the determination of all the functions in terms of a given one.

Example 1:
Express the hyperbolic sine in terms of the other functions.

Solution 1:
From the triangles shown, the following equations are directly obtained:

$$\sinh x = \frac{\sqrt{\cosh^2 x - 1}}{1} \qquad \sinh x = \frac{\sqrt{1 - \mathrm{sech}^2 x}}{\mathrm{sech}\ x}$$

$$\sinh x = \frac{\tanh x}{\sqrt{1 - \tanh^2 x}} \qquad \sinh x = \frac{1}{\mathrm{csch}\ x}$$

$$\sinh x = \frac{1}{\sqrt{\mathrm{ctnh}^2 x - 1}}$$

Example 2:
Express all of the functions in terms of the hyperbolic sine.

Solution 2:
From the hyperbolic sine triangle, the following equations are directly obtained:

$$\cosh x = \frac{\sqrt{1 + \sinh^2 x}}{1} \qquad \mathrm{sech}\ x = \frac{1}{\sqrt{1 + \sinh^2 x}}$$

$$\tanh x = \frac{\sinh x}{\sqrt{1 + \sinh^2 x}} \qquad \mathrm{csch}\ x = \frac{1}{\sinh x}$$

$$\coth x = \frac{\sqrt{1 + \sinh^2 x}}{\sinh x}$$

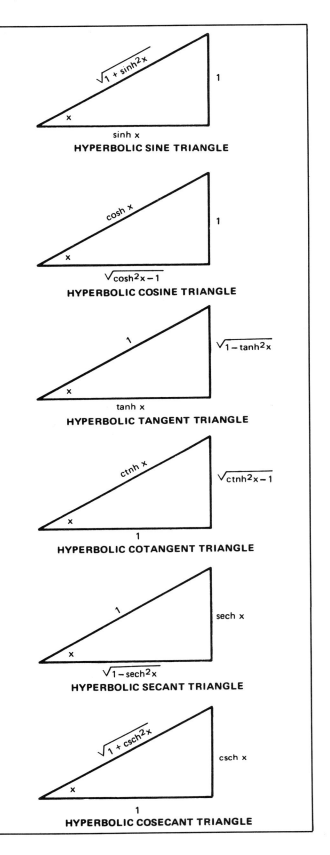

HYPERBOLIC SINE TRIANGLE

HYPERBOLIC COSINE TRIANGLE

HYPERBOLIC TANGENT TRIANGLE

HYPERBOLIC COTANGENT TRIANGLE

HYPERBOLIC SECANT TRIANGLE

HYPERBOLIC COSECANT TRIANGLE

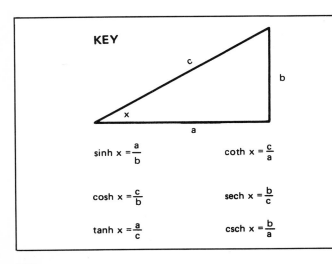

KEY

$$\sinh x = \frac{a}{b} \qquad \coth x = \frac{c}{a}$$

$$\cosh x = \frac{c}{b} \qquad \mathrm{sech}\ x = \frac{b}{c}$$

$$\tanh x = \frac{a}{c} \qquad \mathrm{csch}\ x = \frac{b}{a}$$

Fundamental trigonometric relations

Introduction.
Frequently it is required to determine a fundamental trigonometric relation. When using the following method for determining these relations, the six primary trigonometric functions are placed at the corners of a hexagon. When these functions are placed in a particular order, their fundamental relations can be easily found. The examples that follow the figure explain the usage of this method.

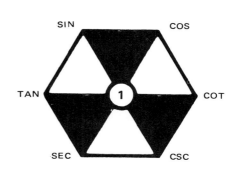

HEXAGON SIDES

CLOCKWISE

$$\tan\theta = \frac{\sin\theta}{\cos\theta}$$

$$\sin\theta = \frac{\cos\theta}{\cot\theta}$$

$$(\tan\theta)(\cos\theta) = \sin\theta$$

$$(\tan\theta)(\csc\theta) = \sec\theta$$

COUNTERCLOCKWISE

$$\cos\theta = \frac{\sin\theta}{\tan\theta}$$

$$\sin\theta = \frac{\tan\theta}{\sec\theta}$$

HEXAGON CORNERS

$$\frac{1}{\sin\theta} = \csc\theta \qquad (\sin\theta)(\csc\theta) = 1$$

$$\frac{1}{\cos\theta} = \sec\theta \qquad (\tan\theta)(\cot\theta) = 1$$

HEXAGON SIDES OF COLORED TRIANGLES

$$\sin^2\theta + \cos^2\theta = 1^2$$

$$\tan^2\theta + 1^2 = \sec^2\theta$$

$$1^2 + \cot^2\theta = \csc^2\theta$$

Polar to rectangular coordinates nomogram

Introduction.
Conversion from polar to rectangular coordinates is simplified by using this nomogram, which consists of nomograms of both systems having the same unit measurement.

Example 1:
Given a polar coordinate system (5 at an angle of 37 deg), determine its rectangular coordinates.

Solution 1:
Move along the positive x-axis until you arrive at the coordinate 5. Follow this circular line coordinate counter-clockwise until it intersects the angle of 37 deg and at this intersection point, read the answer of 4 + J3.

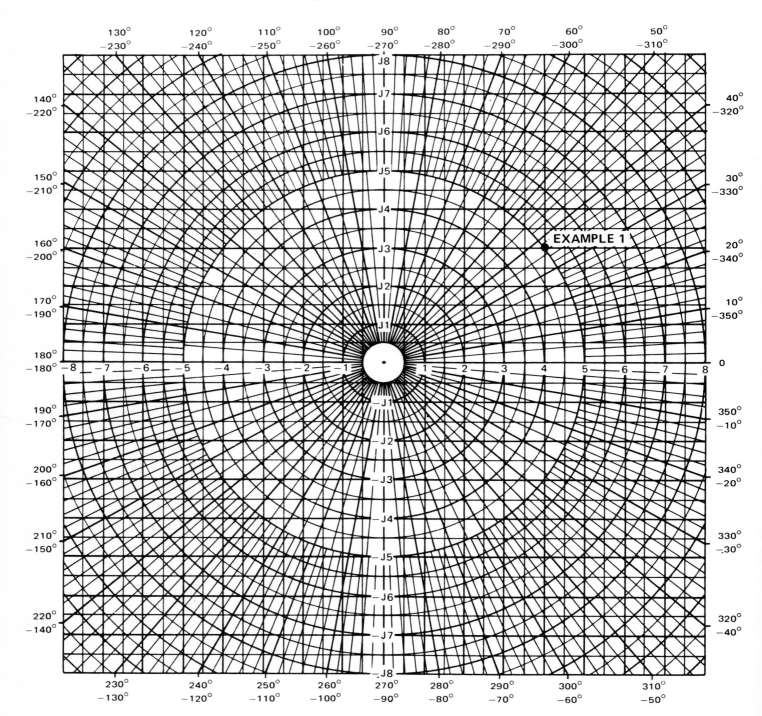

Simplified solution of triangles

Introduction.

Frequently in engineering calculations it is necessary to solve for the angle and/or side of a triangle. The following charts will help to expedite this process.

Example 1:

Given a triangle that has a = 7.0 inches, b = 9.5 inches and angle C = 60°, determine side c.

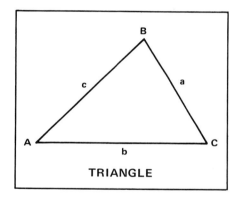

TRIANGLE

Solution 1:

Going to the table we locate the law of cosines:

$$c^2 = a^2 + b^2 - 2ab \cos C \qquad (1)$$

Substituting into Eq. 1:

$c^2 = 7.0^2 + 9.5^2 - 2(7.0)(9.5) \cos 60°$
$c^2 = 49 + 90.2 - 2(7.0)(9.5)(0.500)$
$c^2 = 49 + 90.2 - 66.5$
$c^2 = 72.7$
$c = 8.5$ inches

LAW OF SINES

$$\frac{a}{\sin A} = \frac{b}{\sin B} = \frac{c}{\sin C}$$

LAW OF COSINES

$$c^2 = a^2 + b^2 - 2ab \cos C$$

LAW OF TANGENTS

$$\frac{\tan \frac{1}{2}(A - B)}{\tan \frac{1}{2}(A + B)} = \frac{a - b}{a + b}$$

$$\frac{a}{\sin A} = \frac{b}{\sin B} = \frac{c}{\sin C} = \text{MODULUS}$$

$a = M \sin A$
$b = M \sin B$
$c = M \sin C$

---

3 SIDES KNOWN

FIND: $\cos A = \dfrac{b^2 + c^2 - a^2}{2bc}$ THEN: $\dfrac{a}{\sin A} = \dfrac{b}{\sin B} = \dfrac{c}{\sin C}$

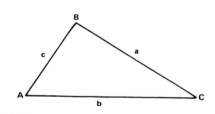

---

2 SIDES AND 1 OPPOSITE ANGLE KNOWN

FIND: $\sin B = \dfrac{b \sin A}{a}$ THEN: $\dfrac{a}{\sin A} = \dfrac{b}{\sin B} = \dfrac{c}{\sin C}$

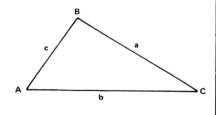

IDENTITIES

$$\frac{\sin x}{\cos x} = \tan x$$

$$\sin^2 x + \cos^2 x = 1$$

AREA OF TRIANGLE

$$\text{AREA} = \sqrt{s(s-a)(s-b)(s-c)}$$

WHERE $S = \dfrac{a + b + c}{2} = (\tfrac{1}{2}\text{sum of 3 sides})$

WHEN AREA (K) OF TRIANGLE IS KNOWN

$\sin A = \dfrac{2K}{bc}$

$\sin B = \dfrac{2K}{ac}$

$\sin C = \dfrac{2K}{ab}$

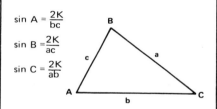

---

2 SIDES AND INCLUDED ANGLE KNOWN

FIND: $\tan B = \dfrac{b \sin A}{c - (b \cos A)}$ THEN: $\dfrac{a}{\sin A} = \dfrac{b}{\sin B} = \dfrac{c}{\sin C}$

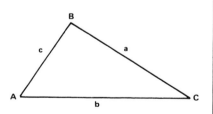

---

1 SIDE AND 2 ANGLES KNOWN

FIND: $B = 180° - (A + C)$ THEN: $\dfrac{a}{\sin A} = \dfrac{b}{\sin B} = \dfrac{c}{\sin C}$

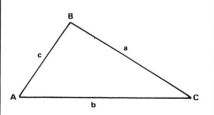

Simplified solution of hyperbolic functions and exponentials

Introduction.
The solution of many mathematical problems can be simplified by use of this procedure for solving hyperbolic functions and exponentials. This procedure is simple and is self-explanatory by merely following the examples below the figure.

SIMPLIFIED PROCEDURE FOR SOLVING HYPERBOLIC FUNCTIONS

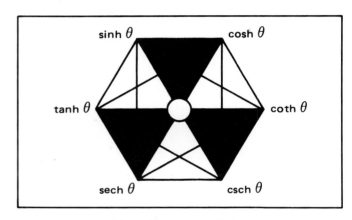

Along Hexagon Sides

Clockwise

$$\tanh\theta = \frac{\sinh\theta}{\cosh\theta} \qquad\qquad \coth\theta = \frac{\operatorname{csch}\theta}{\operatorname{sech}\theta}$$

$$\sinh\theta = \frac{\cosh\theta}{\coth\theta} \qquad\qquad \operatorname{csch}\theta = \frac{\operatorname{sech}\theta}{\tanh\theta}$$

$$\cosh\theta = \frac{\coth\theta}{\operatorname{csch}\theta} \qquad\qquad \operatorname{sech}\theta = \frac{\tanh\theta}{\sinh\theta}$$

Counterclockwise

$$\cosh\theta = \frac{\sinh\theta}{\tanh\theta} \qquad\qquad \operatorname{sech}\theta = \frac{\operatorname{csch}\theta}{\coth\theta}$$

$$\sinh\theta = \frac{\tanh\theta}{\operatorname{sech}\theta} \qquad\qquad \operatorname{csch}\theta = \frac{\coth\theta}{\cosh\theta}$$

$$\tanh\theta = \frac{\operatorname{sech}\theta}{\operatorname{csch}\theta} \qquad\qquad \coth\theta = \frac{\cosh\theta}{\sinh\theta}$$

Along Diagonals (Of Quadrilaterals) Intersecting Hex Corners

$$\tanh\theta\,(\cosh\theta) = \sinh\theta \qquad \coth\theta\,(\operatorname{sech}\theta) = \operatorname{csch}\theta$$
$$\sinh\theta\,(\coth\theta) = \cosh\theta \qquad \operatorname{csch}\theta\,(\tanh\theta) = \operatorname{sech}\theta$$
$$\cosh\theta\,(\operatorname{csch}\theta) = \coth\theta \qquad \operatorname{sech}\theta\,(\sinh\theta) = \tanh\theta$$

Along Lines Across Opposite Hex Corners

$$\operatorname{csch}\theta = \frac{1}{\sinh\theta} \qquad\qquad \sinh\theta\,(\operatorname{csch}\theta) = 1$$

$$\operatorname{sech}\theta = \frac{1}{\cosh\theta} \qquad\qquad \cosh\theta\,(\operatorname{sech}\theta) = 1$$

$$\coth\theta = \frac{1}{\tanh\theta} \qquad\qquad \tanh\theta\,(\coth\theta) = 1$$

Along Sides Of Colored Triangles

Clockwise

$\text{sech}^2\theta + \tanh^2\theta = 1$
$\text{csch}^2\theta + 1 = \coth^2\theta$
$1 + \sinh^2\theta = \cosh^2\theta$

Counterclockwise

$1 - \tanh^2\theta = \text{sech}^2\theta$
$\coth^2\theta - 1 = \text{csch}^2\theta$
$\cosh^2\theta - \sinh^2\theta = 1$

SIMPLIFIED PROCEDURE FOR SOLVING EXPONENTIALS

From the basic fundamental relationships of exponentials to hyperbolic functions which are:

$$\sinh x = \tfrac{1}{2}(e^x - e^{-x}) \qquad \text{csch} x = \frac{2}{e^x - e^{-x}}$$

$$\cosh x = \tfrac{1}{2}(e^x + e^{-x}) \qquad \text{sech} x = \frac{2}{e^x + e^{-x}}$$

$$\tanh x = \frac{e^x - e^{-x}}{e^x + e^{-x}} \qquad \coth x = \left(\frac{e^x + e^{-x}}{e^x - e^{-x}}\right)$$

This simplified procedure can be applied:

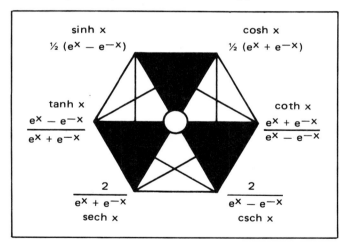

sinh x
$\tfrac{1}{2}(e^x - e^{-x})$

cosh x
$\tfrac{1}{2}(e^x + e^{-x})$

tanh x
$\dfrac{e^x - e^{-x}}{e^x + e^{-x}}$

coth x
$\dfrac{e^x + e^{-x}}{e^x - e^{-x}}$

$\dfrac{2}{e^x + e^{-x}}$
sech x

$\dfrac{2}{e^x - e^{-x}}$
csch x

Along Hexagon Sides

Clockwise

$$\tanh x = \frac{\tfrac{1}{2}(e^x - e^{-x})}{\tfrac{1}{2}(e^x + e^{-x})} = \frac{e^x - e^{-x}}{e^x + e^{-x}}$$

$$\sinh x = \frac{\tfrac{1}{2}(e^x + e^{-x})}{\left(\dfrac{e^x + e^{-x}}{e^x - e^{-x}}\right)} = \frac{e^x - e^{-x}}{2}$$

$$\cosh x = \frac{\left(\dfrac{e^x + e^{-x}}{e^x - e^{-x}}\right)}{\dfrac{2}{e^x - e^{-x}}} = \frac{e^x + e^{-x}}{2}$$

$$\coth x = \frac{\left(\dfrac{2}{e^x - e^{-x}}\right)}{\left(\dfrac{2}{e^x + e^{-x}}\right)} = \frac{e^x + e^{-x}}{e^x - e^{-x}}$$

$$\text{csch} x = \frac{\left(\dfrac{2}{e^x + e^{-x}}\right)}{\left(\dfrac{e^x - e^{-x}}{e^x + e^{-x}}\right)} = \frac{2}{(e^x - e^{-x})}$$

$$\text{sech} x = \frac{\left(\dfrac{e^x - e^{-x}}{e^x + e^{-x}}\right)}{\left(\dfrac{e^x - e^{-x}}{2}\right)} = \frac{2}{e^x + e^{-x}}$$

Counterclockwise

$$\cosh x = \frac{\left(\dfrac{e^x - e^{-x}}{2}\right)}{\left(\dfrac{e^x - e^{-x}}{e^x + e^{-x}}\right)} = \frac{e^x + e^{-x}}{2}$$

$$\sinh x = \frac{\left(\dfrac{e^x - e^{-x}}{e^x + e^{-x}}\right)}{\left(\dfrac{2}{e^x + e^{-x}}\right)} = \frac{e^x - e^{-x}}{2}$$

$$\tanh x = \frac{\left(\dfrac{2}{e^x + e^{-x}}\right)}{\left(\dfrac{2}{e^x - e^{-x}}\right)} = \frac{e^x - e^{-x}}{e^x + e^{-x}}$$

$$\text{sech} x = \frac{\left(\dfrac{2}{e^x - e^{-x}}\right)}{\left(\dfrac{e^x + e^{-x}}{e^x - e^{-x}}\right)} = \frac{2}{(e^x + e^{-x})}$$

$$\text{csch} x = \frac{\dfrac{e^x + e^{-x}}{e^x - e^{-x}}}{\dfrac{e^x + e^{-x}}{2}} = \frac{2}{e^x - e^{-x}}$$

$$\coth x = \frac{\dfrac{e^x + e^{-x}}{2}}{\dfrac{e^x - e^{-x}}{2}} = \frac{e^x + e^{-x}}{e^x - e^{-x}}$$

Along Diagonals (Of Quadrilaterals) Intersecting Hex Corners

$$\text{sinh}x = \frac{e^x - e^{-x}}{e^x + e^{-x}}\left(\frac{e^x + e^{-x}}{2}\right) = \frac{e^x - e^{-x}}{2}$$

$$\text{cosh}x = \frac{e^x - e^{-x}}{2}\left(\frac{e^x + e^{-x}}{e^x - e^{-x}}\right) = \frac{e^x + e^{-x}}{2}$$

$$\text{coth}x = \frac{e^x + e^{-x}}{2}\left(\frac{2}{e^x - e^{-x}}\right) = \frac{e^x + e^{-x}}{e^x - e^{-x}}$$

$$\text{csch}x = \frac{e^x + e^{-x}}{e^x - e^{-x}}\left(\frac{2}{e^x + e^{-x}}\right) = \frac{2}{e^x - e^{-x}}$$

$$\text{sech}x = \frac{2}{e^x - e^{-x}}\left(\frac{e^x - e^{-x}}{e^x + e^{-x}}\right) = \frac{2}{e^x + e^{-x}}$$

$$\text{tanh}x = \frac{2}{e^x + e^{-x}}\left(\frac{e^x - e^{-x}}{2}\right) = \frac{e^x - e^{-x}}{e^x + e^{-x}}$$

Along Lines Across Opposite Hex Corners

$$\text{csch}x = \frac{1}{\frac{1}{2}(e^x - e^{-x})} = \frac{2}{e^x - e^{-x}}$$

$$\text{sech}x = \frac{1}{\frac{1}{2}(e^x + e^{-x})} = \frac{2}{e^x + e^{-x}}$$

$$\text{coth}x = \frac{1}{\frac{e^x - e^{-x}}{e^x + e^{-x}}} = \frac{e^x + e^{-x}}{e^x - e^{-x}}$$

or

$$\text{sinh}x\,(\text{csch}x) = 1, \left(\frac{e^x - e^{-x}}{2}\right)\left(\frac{2}{e^x - e^{-x}}\right) = 1$$

$$\text{cosh}x\,(\text{sech}x) = 1, \left(\frac{e^x + e^{-x}}{2}\right)\left(\frac{2}{e^x + e^{-x}}\right) = 1$$

$$\text{tanh}x\,(\text{coth}x) = 1, \left(\frac{e^x - e^{-x}}{e^x + e^{-x}}\right)\left(\frac{e^x + e^{-x}}{e^x - e^{-x}}\right) = 1$$

Along Sides Of Colored Triangles Clockwise

$$\text{sech}^2x + \text{tanh}^2x = 1$$

$$\left(\frac{2}{e^x + e^{-x}}\right)^2 + \left(\frac{e^x - e^{-x}}{e^x + e^{-x}}\right)^2 = 1$$

$$\text{csch}^2x + 1 = \text{coth}^2x$$

$$\left(\frac{2}{e^x - e^{-x}}\right)^2 + 1 = \left(\frac{e^x + e^{-x}}{e^x - e^{-x}}\right)^2$$

$$1 + \text{sinh}^2x = \text{cosh}^2x$$

$$1 + \left(\frac{e^x - e^{-x}}{2}\right)^2 = \left(\frac{e^x + e^{-x}}{2}\right)^2$$

Counterclockwise

$$1 - \text{tanh}^2x = \text{sech}^2x$$

$$1 - \left(\frac{e^x - e^{-x}}{e^x + e^{-x}}\right)^2 = \left(\frac{2}{e^x + e^{-x}}\right)^2$$

$$\text{cosh}^2x - \text{sinh}^2x = 1$$

$$\left(\frac{e^x + e^{-x}}{2}\right)^2 - \left(\frac{e^x - e^{-x}}{2}\right)^2 = 1$$

$$\text{coth}^2x - 1 = \text{csch}^2x$$

$$\left(\frac{e^x + e^{-x}}{e^x - e^{-x}}\right)^2 - 1 = \left(\frac{2}{e^x - e^{-x}}\right)^2$$

Trigonometric functions

Introduction.
Frequently in engineering calculations it is necessary to obtain the sine, cosine and tangent of an angle. The following table expedites that process.

Example 1:
Given the angle of 60°, determine its sine, cosine and tangent.

Solution 1:
Going to the table we arrive at the following answers:

sine 60° = 0.866
cosine 60° = 0.500
tangent 60° = 1.732

Degrees	Sine	Cosine	Tangent	Degrees	Sine	Cosine	Tangent
0	0.000	1.000	0.000				
1	0.017	1.000	0.017	46	0.719	0.695	1.036
2	0.035	0.999	0.035	47	0.731	0.682	1.072
3	0.052	0.999	0.052	48	0.743	0.669	1.111
4	0.070	0.998	0.070	49	0.755	0.656	1.150
5	0.087	0.996	0.087	50	0.766	0.643	1.192
6	0.105	0.995	0.105	51	0.777	0.629	1.235
7	0.122	0.993	0.123	52	0.788	0.616	1.280
8	0.139	0.990	0.141	53	0.799	0.602	1.327
9	0.156	0.988	0.158	54	0.809	0.588	1.376
10	0.174	0.985	0.176	55	0.819	0.574	1.428
11	0.191	0.982	0.194	56	0.829	0.559	1.483
12	0.208	0.978	0.213	57	0.839	0.545	1.540
13	0.225	0.974	0.231	58	0.848	0.530	1.600
14	0.242	0.970	0.249	59	0.857	0.515	1.664
15	0.259	0.966	0.268	60	0.866	0.500	1.732
16	0.276	0.961	0.287	61	0.875	0.485	1.804
17	0.292	0.956	0.306	62	0.883	0.469	1.881
18	0.309	0.951	0.325	63	0.891	0.454	1.963
19	0.326	0.946	0.344	64	0.899	0.438	2.050
20	0.342	0.940	0.364	65	0.906	0.423	2.145
21	0.358	0.934	0.384	66	0.914	0.407	2.246
22	0.375	0.927	0.404	67	0.921	0.391	2.356
23	0.391	0.921	0.424	68	0.927	0.375	2.475
24	0.407	0.914	0.445	69	0.934	0.358	2.605
25	0.423	0.906	0.466	70	0.940	0.342	2.748
26	0.438	0.899	0.488	71	0.946	0.326	2.904
27	0.454	0.891	0.510	72	0.951	0.309	3.078
28	0.469	0.883	0.532	73	0.956	0.292	3.271
29	0.485	0.875	0.554	74	0.961	0.276	3.487
30	0.500	0.866	0.577	75	0.966	0.259	3.732
31	0.515	0.857	0.601	76	0.970	0.242	4.011
32	0.530	0.848	0.625	77	0.974	0.225	4.332
33	0.545	0.839	0.649	78	0.978	0.208	4.705
34	0.559	0.829	0.675	79	0.982	0.191	5.145
35	0.574	0.819	0.700	80	0.985	0.174	5.671
36	0.588	0.809	0.727	81	0.988	0.156	6.314
37	0.602	0.799	0.754	82	0.990	0.139	7.115
38	0.616	0.788	0.781	83	0.993	0.122	8.144
39	0.629	0.777	0.810	84	0.995	0.105	9.514
40	0.643	0.766	0.839	85	0.996	0.087	11.43
41	0.656	0.755	0.869	86	0.998	0.070	14.30
42	0.669	0.743	0.900	87	0.999	0.052	19.08
43	0.682	0.731	0.933	88	0.999	0.035	28.61
44	0.695	0.719	0.966	89	1.000	0.017	57.29
45	0.707	0.707	1.000	90	1.000	0.000	

Tapers, slopes and their angles

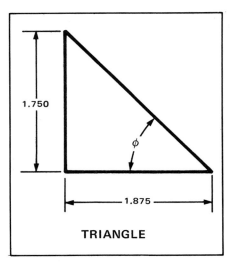

1.750

1.875

TRIANGLE

Introduction.
Frequently in engineering calculations it is necessary to determine the angles of tapers. The following table provides a rapid means of determining the angles of many frequently used tapers and slopes.

Note.
If tan ϕ is greater than 1, interchange the numerator and the denominator and read $90° - \phi$ from the table.

Example 1:
Given a component that has a base of 1.875 and height of 1.750, determine the angle of its taper.

Solution 1:
Solving for the tan ϕ:

$$\tan \phi = \frac{1.750}{1.875} = \frac{14}{15}$$

Locate the taper of 14/15 in the table and read the answer of $\phi = 43° \ 1'30''$.

TAPER	DEGREES	MINUTES	SECONDS	TAPER	DEGREES	MINUTES	SECONDS
1/2	26	33	54	7/13	28	18	3
1/3	18	26	6	8/13	31	36	27
2/3	33	41	24	9/13	34	41	43
1/4	14	2	10	10/13	37	34	7
3/4	36	52	12	11/13	40	14	11
1/5	11	18	36	12/13	42	42	34
2/5	21	48	5	1/14	4	5	8
3/5	30	57	50	3/14	12	5	41
4/5	38	39	35	5/14	19	39	14
1/6	9	27	44	9/14	32	44	7
5/6	39	48	20	11/14	38	9	26
1/7	8	7	48	13/14	42	52	44
2/7	15	56	43	1/15	3	48	51
3/7	23	11	55	2/15	7	35	41
4/7	29	44	42	4/15	14	55	53
5/7	35	32	16	7/15	25	1	1
6/7	40	36	5	8/15	28	4	21
1/8	7	7	30	11/15	36	15	14
3/8	20	33	22	13/15	40	54	52
5/8	32	0	19	14/15	43	1	30
7/8	41	11	9	1/16	3	34	35
1/9	6	20	25	3/16	10	37	11
2/9	12	31	44	5/16	17	21	14
4/9	23	57	45	7/16	23	37	46
5/9	29	3	17	9/16	29	21	28
7/9	37	52	30	11/16	34	30	31
8/9	41	38	1	13/16	39	5	38
1/10	5	42	38	15/16	43	9	9
3/10	16	41	57	1/17	3	21	59
7/10	34	59	31	2/17	6	42	35
9/10	41	59	14	3/17	10	0	29
1/11	5	11	40	4/17	13	14	26
2/11	10	18	17	5/17	16	23	22
3/11	15	15	18	6/17	19	26	24
4/11	19	58	59	7/17	22	22	48
5/11	24	26	38	8/17	25	12	4
6/11	28	36	38	9/17	27	53	50
7/11	32	28	16	10/17	30	27	56
8/11	36	1	39	11/17	32	54	19
9/11	39	17	22	12/17	35	13	3
10/11	42	16	25	13/17	37	24	19
1/12	4	45	49	14/17	39	28	21
5/12	22	37	12	15/17	41	25	25
7/12	30	15	23	16/17	43	15	51
11/12	42	30	38	1/18	3	10	47
1/13	4	23	55	5/18	15	31	27
2/13	8	44	56	7/18	21	15	2
3/13	12	59	41	11/18	31	25	46
4/13	17	6	10	13/18	35	50	16
5/13	21	2	15	17/18	43	21	48
6/13	24	46	31	1/19	3	0	46

TAPER	DEGREES	MINUTES	SECONDS	TAPER	DEGREES	MINUTES	SECONDS
2/19	6	0	32	3/23	7	25	53
3/19	8	58	21	4/23	9	51	57
4/19	11	53	19	5/23	12	15	53
5/19	14	44	37	6/23	14	37	15
6/19	17	31	32	7/23	16	55	39
7/19	20	13	30	8/23	19	10	44
8/19	22	50	1	9/23	21	22	14
9/19	25	20	46	10/23	23	29	55
10/19	27	45	31	11/23	25	33	36
11/19	30	4	7	12/23	27	33	10
12/19	32	16	32	13/23	29	28	33
13/19	34	22	49	14/23	31	19	43
14/19	36	23	4	15/23	33	6	41
15/19	38	17	25	16/23	34	49	28
16/19	40	6	3	17/23	36	28	9
17/19	41	49	13	18/23	38	2	49
18/19	43	27	7	19/23	39	33	35
1/20	2	51	45	20/23	41	0	33
3/20	8	31	51	21/23	42	23	51
7/20	19	17	24	22/23	43	43	37
9/20	24	13	40	1/24	2	23	9
11/20	28	48	39	5/24	11	46	6
13/20	33	1	26	7/24	16	15	37
17/20	40	21	52	11/24	24	37	25
19/20	43	31	52	13/24	28	26	35
1/21	2	43	35	17/24	35	18	40
2/21	5	26	25	19/24	38	22	3
4/21	10	47	3	23/24	43	46	52
5/21	13	23	33	1/25	2	17	26
8/21	20	51	16	2/25	4	34	26
10/21	25	27	48	3/25	6	50	34
11/21	27	38	46	4/25	9	5	25
13/21	31	45	34	6/25	13	29	45
16/21	37	18	14	7/25	15	38	32
17/21	38	59	28	8/25	17	44	41
19/21	42	8	15	9/25	19	47	56
20/21	43	36	10	11/25	23	44	58
1/22	2	36	9	12/25	25	38	28
3/22	7	45	55	13/25	27	28	28
5/22	12	48	15	14/25	29	14	56
7/22	17	39	1	16/25	32	37	9
9/22	22	14	56	17/25	34	12	57
13/22	30	34	45	18/25	35	45	14
15/22	34	17	13	19/25	37	14	5
17/22	37	41	39	21/25	40	1	49
19/22	40	48	54	22/25	41	20	52
21/22	43	40	4	23/25	42	36	51
1/23	2	29	22	24/25	43	49	51
2/23	4	58	11				

Development of a symmetrical cone

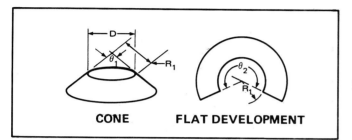

CONE   **FLAT DEVELOPMENT**

Introduction.

On occasion, the engineer is faced with a problem of developing three-dimensional objects from flat sheet. The required flat layout for developing cones is simplified by use of the following table that is based on these equations:

$$\theta_2 = \frac{360D}{2R_1} \tag{1}$$

$$\theta_2 = \frac{180D}{R_1} \tag{2}$$

Nomenclature:

θ_1 = Apex angle, deg

θ_2 = Cone developed angle in flat sheet, deg

D = Diameter at cone top, inches

R_1 = Radius from developed cone top to apex, inches

The diameter at the top of the developed cone is found from the following equation:

$$D = \sqrt{2R_1^2 - [2R_1^2 \cos \theta_1]} \tag{3}$$

Note.

The following table gives the exact angle required without having to make use of Eqs. 1 or 2. Keep in mind that it doesn't allow for laps or joining of the cone after it has been formed. Material must be provided for tongues or overlaps.

Example 1:

Develop a cone having a θ_1 = 30 deg and radius from the top of the developed cone to the apex equal to 4.0 inches.

Solution 1:

Substituting into Eq. 3 we arrive at:

$$D = \sqrt{(2)\,(4.0^2) - [(2)\,(4.0^2)\,(0.8660)]}$$

$$D = \sqrt{32 - 27.7}$$

$$D = \sqrt{4.3}$$

$$D = 2.073 \text{ inches}$$

Going to the table we locate θ_1 = 30 deg and observe that $\theta_2 = 93.17°$.

CONE θ_1	FLAT θ_2	CONE θ_1	FLAT θ_2	CONE θ_1	FLAT θ_2
1	3.1415	61	182.714	121	313.328
2	6.2829	62	185.414	122	314.863
3	9.4237	63	188.100	123	316.374
4	12.5638	64	190.771	124	317.861
5	15.7030	65	193.428	125	319.324
6	18.8410	66	196.070	126	320.762
7	21.9775	67	198.697	127	322.177
8	25.1124	68	201.310	128	323.566
9	28.2453	69	203.906	129	324.931
10	31.3761	70	206.488	130	326.271
11	34.5045	71	209.053	131	327.586
12	37.6303	72	211.603	132	328.876
13	40.7532	73	214.136	133	330.142
14	43.8730	74	216.654	134	331.382
15	46.9895	75	219.154	135	332.597
16	50.1024	76	221.638	136	333.786
17	53.2114	77	224.105	137	334.950
18	56.3164	78	226.555	138	336.089
19	59.4172	79	228.988	139	337.202
20	62.5134	80	231.404	140	338.289
21	65.6048	81	233.801	141	339.351
22	68.6913	82	236.181	142	340.387
23	71.7725	83	238.543	143	341.397
24	74.8483	84	240.887	144	342.380
25	77.9183	85	243.213	145	343.338
26	80.9824	86	245.520	146	344.270
27	84.0404	87	247.808	147	345.175
28	87.0919	88	250.077	148	346.054
29	90.1369	89	252.327	149	346.907
30	93.1749	90	254.559	150	347.733
31	96.2059	91	256.770	151	348.533
32	99.2295	92	258.962	152	349.307
33	102.246	93	261.135	153	350.053
34	105.254	94	263.287	154	350.773
35	108.254	95	265.420	155	351.467
36	111.246	96	267.532	156	352.133
37	114.230	97	269.624	157	352.773
38	117.205	98	271.696	158	353.386
39	120.171	99	273.746	159	353.972
40	123.127	100	275.776	160	354.531
41	126.075	101	277.785	161	355.063
42	129.013	102	279.773	162	355.568
43	131.941	103	281.739	163	356.046
44	134.858	104	283.684	164	356.497
45	137.766	105	285.607	165	356.920
46	140.663	106	287.509	166	357.317
47	143.550	107	289.389	167	357.686
48	146.425	108	291.246	168	358.028
49	149.290	109	293.082	169	358.343
50	152.143	110	294.895	170	358.630
51	154.984	111	296.686	171	358.890
52	157.814	1.12	298.454	172	359.123
53	160.631	113	300.199	173	359.329
54	163.437	114	301.922	174	359.507
55	166.230	115	303.621	175	359.657
56	169.010	116	305.297	176	359.781
57	171.777	117	306.951	177	359.877
58	174.532	118	308.580	178	359.945
59	177.273	119	310.187	179	359.986
60	180.000	120	311.769		

REQUIRED ANGLE FOR CONE DEVELOPMENT

Numerical solution
of complex equations

Introduction.

Sometimes it is necessary to solve higher-degree equations and equations containing trigonometric or logarithmic terms. The roots, or values of x that will satisfy such equations, must sometimes be determined to a high degree of accuracy.

As an aid in finding a solution, the equation can be set equal to ordinate y, thus permitting a two-dimensional plot to be made. This has been done in a general way in Fig. 1. The curve crosses the x-axis at point A and ordinate y at this point is equal to zero. Hence, the x-value of point A will satisfy the given equation—which is also equal to zero and is therefore a root.

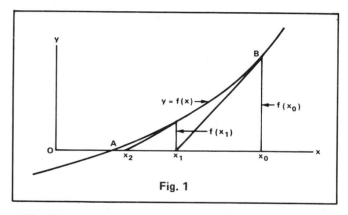

Fig. 1

The Newton-Raphson method of solution consists of first estimating a root from the plot of the equation and then improving on it by arithmetic calculations. Suppose in Fig. 1 that the engineer estimates a root has a value x_0. This is obviously not a very good estimate but it will serve to illustrate the process.

The ordinate $f(x_0)$ can be calculated from the equation. Let the tangent $\frac{dy}{dx}$ or $f'(x_0)$ to the curve be drawn at B and let it extend until it intersects the x-axis. It is obvious that x_1 is a better estimate for root A than the first estimate x_0. The slope or tangent $f'(x_0)$ at point x_0 is equal to:

$$f'(x_0) = \frac{f(x_0)}{x_0 - x_1}$$

or

$$x_1 = x_0 - \frac{f(x_0)}{f'(x_0)} \tag{1}$$

This process can be continued using x_1 as a new initial value and we arrive at the following equation:

$$x_2 = x_1 - \frac{f(x_1)}{f'(x_1)} \tag{2}$$

It is obvious that x_2 is closer to the true root than x_1. The process can be continued until the root is determined as closely as desired.

Example 1:
Given the equation $x^3 - 8x^2 + 18x - 10 = 0$, determine the smallest of the three roots.

Solution 1:
Set the equation equal to y and plot as in Fig. 2:

$$y = f(x) = x^3 - 8x^2 + 18x - 10 \qquad (3)$$

To find the equation of the slope, the equation should be differentiated:

$$\frac{dy}{dx} = f'(x) = 3x^2 - 16x + 18 \qquad (4)$$

From Fig. 2 it appears the lowest root can be estimated as approximately equal to 0.9. Thus from Eq. 3:

$$f(x) = 0.9^3 - (8)(0.9^2) + 18(0.9) - 10$$
$$f(x) = 0.449$$

From Eq. 4:

$$f'(x) = (3)(0.9^2) - (16)(0.9) + 18$$
$$f'(x) = 6.03$$

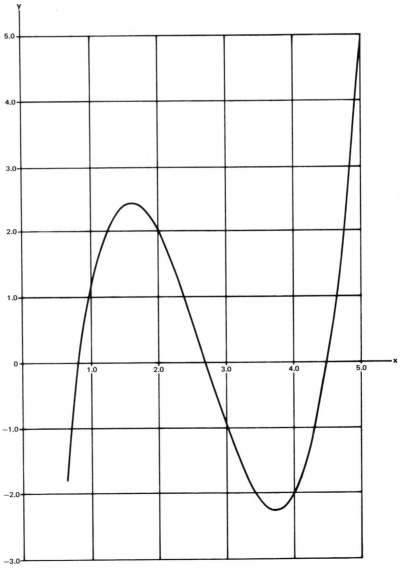

Fig. 2

136

From Eq. 1:

$$x_1 = 0.9 - \frac{0.449}{6.03}$$
$$x_1 = 0.9 - 0.0745$$
$$x_1 = 0.8255$$

This value can be improved by repeating the above process and using Eq. 2 instead of Eq. 1.

From Eq. 3:

$$f(x_1) = 0.8255^3 - (8)(0.8255^2) + (18)(0.8255) - 10$$
$$f(x_1) = -0.0301$$

From Eq. 4:

$$f'(x_1) = (3)(0.8255^2) - (16)(0.8255) + 18$$
$$f'(x_1) = 6.8364$$

From Eq. 2:

$$x_2 = 0.8255 - \frac{-0.0301}{6.8364}$$
$$x_2 = 0.8299$$

Further calculations will indicate no further change in x. In fact, this result is accurate to four decimals.

Note.

The same process can be applied to finding the middle and the largest root of Eq. 1. From Fig. 2, the estimate for the middle root is 2.7 and the largest root is 4.5.

Care must be exercised that the slope does not become zero at some point between A and B and the curve does not have a point of inflection in this interval.

Example 2:

Given the equation $2x + \sin x - 3.306 = 0$, determine its root.

Solution 2:

Set the equation equal to y and plot as in Fig. 3:

$$y = f(x) = 2x + \sin x - 3.306 \qquad (5)$$

To find the equation of the slope, the equation should be differentiated.

$$\frac{dy}{dx} = f'(x) = 2 + \cos x \qquad (6)$$

From Fig. 3 it appears the root can be estimated as approximately equal to 1.200. The variable x is in radian measure and a table showing the trigonometric functions for radians should be used.

From Eq. 5:

$$f(x) = (2)(1.200) + 0.932 - 3.306$$
$$f(x) = 0.026$$

From Eq. 6:

$$f'(x) = 2 + 0.362$$
$$f'(x) = 2.362$$

From Eq. 1:

$$x_1 = 1.200 - \frac{0.026}{2.362}$$
$$x_1 = 1.200 - 0.011$$
$$x_1 = 1.189$$

That this is a sufficiently accurate solution can be shown by substituting into Eq. 5:

$$f(x) = (2)(1.189) + 0.928 - 3.306 = 0$$

Algebraic methods exist for the solution of general cubic and quartic equations but these are frequently very laborious to apply. The numerical method illustrated in this article can be done far more easily and in general will give satisfactory results.

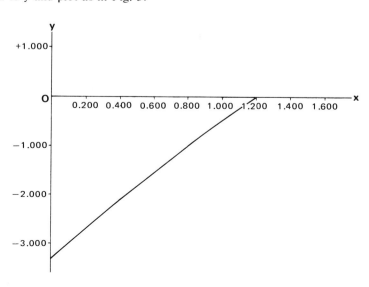

Fig. 3

Solutions for the 'unsolvable' equations

Introduction.

Frequently an engineer is faced with the problem of solving an unsolvable equation. This usually happens when a design problem involves a differential equation and after integrating this equation, the variable x occurs in a mathematical expression or equation, unsolvable by ordinary methods.

There may be a number of types of unsolvable equations. For example:

$$y = f(x) + g(x) \tag{A}$$
$$y = f(x) \cdot g(x) \tag{B}$$
$$y = x \cdot f(x) \tag{C}$$

where f(x) and g(x) mean any function of x.

If the numerical value of y is known, there are two ways of finding x from the above type of equations. Both of these methods (graphical and numerical) are long and troublesome.

Graphical Method.

By this method, a graph of a given function is drawn and the proper value of x is found graphically.

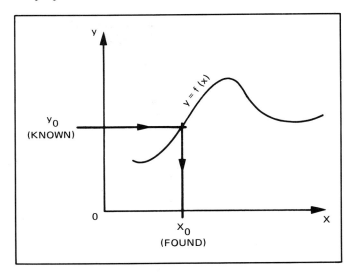

Numerical Method.

This method, also known as the trial-and-error method, is based on selecting a random value of x, solving the equation for y, and comparing the obtained result with the given value for y_0.

If the value found is far away from the given value, change the x value in a desirable direction. Repeat this operation, approaching the known value of y, until the desired value of x is found within satisfactory accuracy.

Among the numerous possible unsolvable equations, the following type is the one that occurs most often:

$$y = x \cdot f(x) \tag{D}$$

The following five basic equations of this type have been chosen for study in this article:

$$y = x^x \tag{1}$$
$$y = x \cdot e^x \tag{2}$$
$$y = x \cdot \log x \tag{3}$$
$$y = x \cdot \sin x \tag{4}$$
$$y = x \cdot \tan x \tag{5}$$

For each of the above equations, a nomogram has been constructed. When the numerical value of y in the particular equation is known, the x value is quickly found.

Use of Nomograms.

Although the graphical forms of the five nomograms are different, their construction principal is identical. Each nomogram consists of a single scale for y and two scales for x (called primary and secondary scales).

The method of using the five nomograms is identical. One end of a straight edge is placed on the known value of y; the other end is moved so that it intersects both x scales (primary and secondary) on identical values of x. This value of x is the solution of the given equation.

In some cases, there are two values of x necessary to solve the given equation.

Nomogram 1.

This nomogram was constructed for solving Eq. 1.

$$y = x^x$$

The nomogram is divided into the following two ranges:

range 1: $y \leq 1$
range 2: $y \geq 1$

Also, x has to be positive since the function $y = x^x$ for negative values of x is not continuous and often has no meaning. The equation $y = x^x$ in range 1 always has two solutions and in range 2, has only one solution.

Example 1:

Determine x from the following equation:

$$0.80 = x^x$$

Solution 1:

In range 1 of nomogram 1, read the following values:

$x_1 = 0.095$
$x_2 = 0.74$

Example 2:

Determine x from the following equation:

$$100 = x^x$$

Solution 2:

In range 2 of nomogram 1, read the following value:

$x_1 = 3.6$

Nomogram 2.

This nomogram was constructed for solving Eq. 2.

$$y = x \cdot e^x$$

The nomogram is divided into the following two ranges:

range 1: $y < e$
range 2: $y > e$

The letter e is the basis of the natural logarithm and is approximately equal to 2.718.

Example 3:

Determine x from the following equation:

$$1.0 = x \cdot e^x$$

Solution 3:

In range 1 of nomogram 2, read the following value:

$x_1 = 0.565$

Example 4:

Determine x from the following equation:

$$100 = x \cdot e^x$$

Solution 4:

In range 2 of nomogram 2, read the following value:

$x_2 = 3.39$

Nomogram 3.

This nomogram was constructed for solving Eq. 3.

$$y = x \cdot \log x$$

The nomogram is divided into the following two ranges:

range 1: $y < 0$
range 2: $y > 0$

Because of the characteristics of Eq. 3 in the two ranges of nomogram 3, the x values are exclusively positive. Eq. 3 has no meaning for the values $x < 0$. For negative y values, $y < 0$, the equation always has two solutions while the same equation for positive y values, $y > 0$, always has only one solution.

Since the decimal logarithm can be converted to a natural logarithm by means of the following factor:

$$\ln x = 2.303 \log_{10} x$$

Nomogram 3 may also be used to solve the following equation:

$$y = x \cdot \ln x$$

For example, it may be necessary to solve the following equation:

$$10 = x \cdot \ln x$$

This equation may be converted to:

$$x \cdot \log x = \frac{10}{2.303}$$

$$x \cdot \log x = 4.324$$

Now in this form, it may be solved by nomogram 3.

Example 5:
Determine x from the following equation:

$$-0.12 = x \cdot \log x$$

Solution 5:
In range 1 of nomogram 3, read the following values:

$$x_1 = 0.14$$
$$x_2 = 0.65$$

Example 6:
Determine x from the following equation:

$$100 = x \cdot \log x$$

Solution 6:
In range 2 of nomogram 3, read the following value:

$$x_1 = 57$$

Nomogram 4.
This nomogram was constructed for solving Eq. 4.

$$y = x \cdot \sin x$$

The nomogram contains the following range:

0-3.1416 radians (0-180°)

The x value is expressed in radians where a radian is approximately equal to 57.296°. The range covered by this nomogram is equal to the first two quadrants of a trigonometric circle. The equation in this range always has two solutions. One solution is always smaller than $1.5708(\frac{\pi}{2})$ and the other always larger than $1.5708(\frac{\pi}{2})$.

Example 7:
Determine x from the following equation:

$$0.50 = x \cdot \sin x$$

Solution 7:
On nomogram 4 read the following values:

$$x_1 = 0.739 \text{ radians}$$
$$x_2 = 2.97 \text{ radians}$$

Nomogram 5.
This nomogram was constructed for solving Eq. 5.

$$y = x \cdot \tan x$$

The nomogram contains the following range:

0-1.5708 radians (0-90°)

The x value is expressed in radians. The range covered by this nomogram is equal to the first quadrant of a trigonometric circle. The equation in this range has only one solution.

Example 8:
Determine x from the following equation:

$$1.0 = x \cdot \tan x$$

Solution 8:
On nomogram 5 read the following value:

$$x = 0.86 \text{ radians}$$

Note.
Nomograms 1 through 5 may also be used to solve ordinary equations containing three unknown values. Examples of such equations are:

$$y = x^z \quad \text{(Nomogram 1)}$$
$$y = x \cdot e^z \quad \text{(Nomogram 2)}$$
$$y = x \cdot \log z \quad \text{(Nomogram 3)}$$
$$y = x \cdot \sin z \quad \text{(Nomogram 4)}$$
$$y = x \cdot \tan z \quad \text{(Nomogram 5)}$$

In all these cases, the y value is unknown and the x and z values are known.

When using the nomograms in this way, the primary x scale stays for x and the secondary x scale stays for z. To use the nomograms, connect the x and z values with a straight edge and read the y value on the intersection point of the straight edge with the y scale.

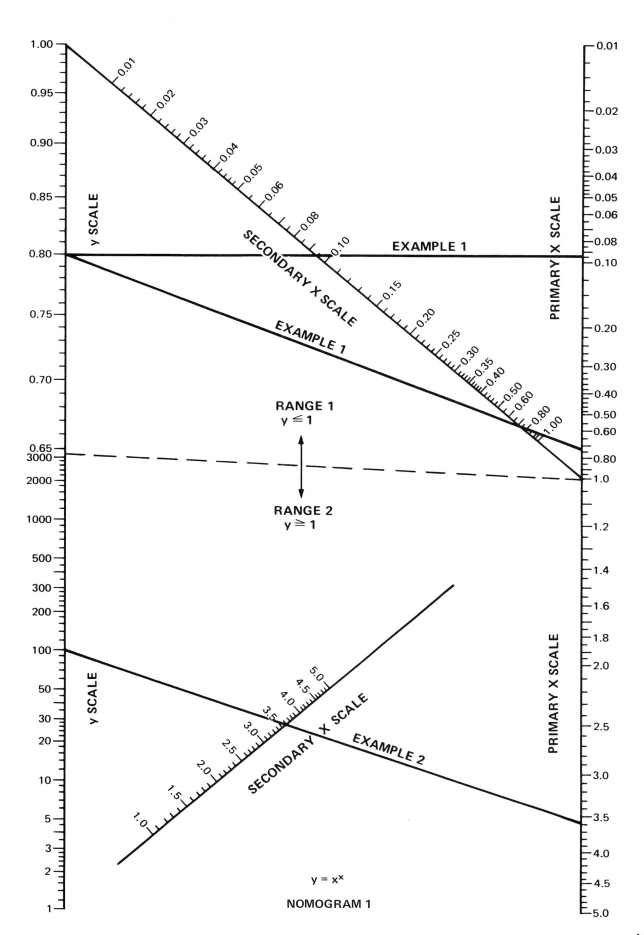

$$y = x^x$$

NOMOGRAM 1

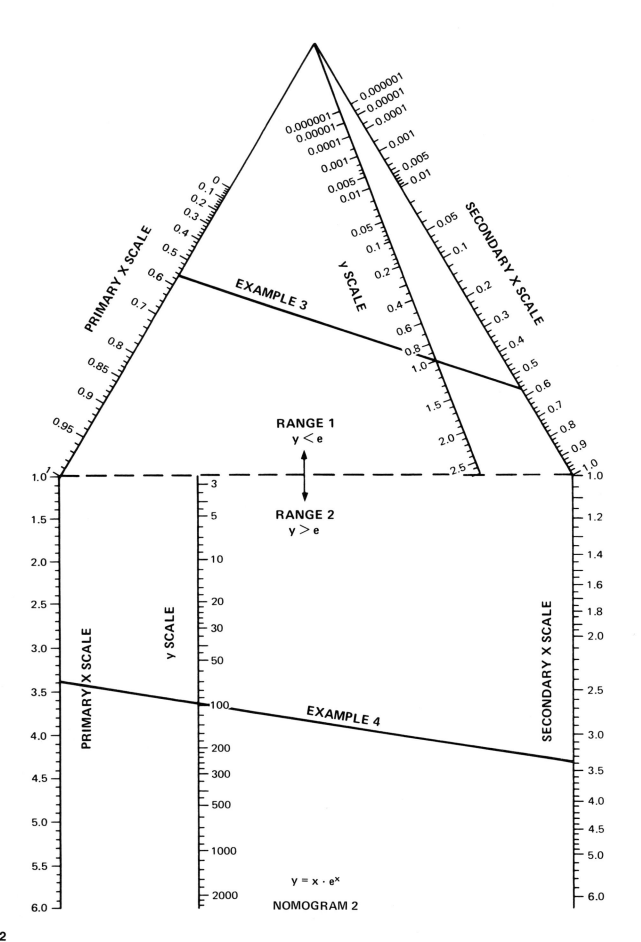

PRIMARY X SCALE

SECONDARY X SCALE

y SCALE

EXAMPLE 3

RANGE 1
y < e

RANGE 2
y > e

PRIMARY X SCALE

y SCALE

SECONDARY X SCALE

EXAMPLE 4

$y = x \cdot e^x$

NOMOGRAM 2

142

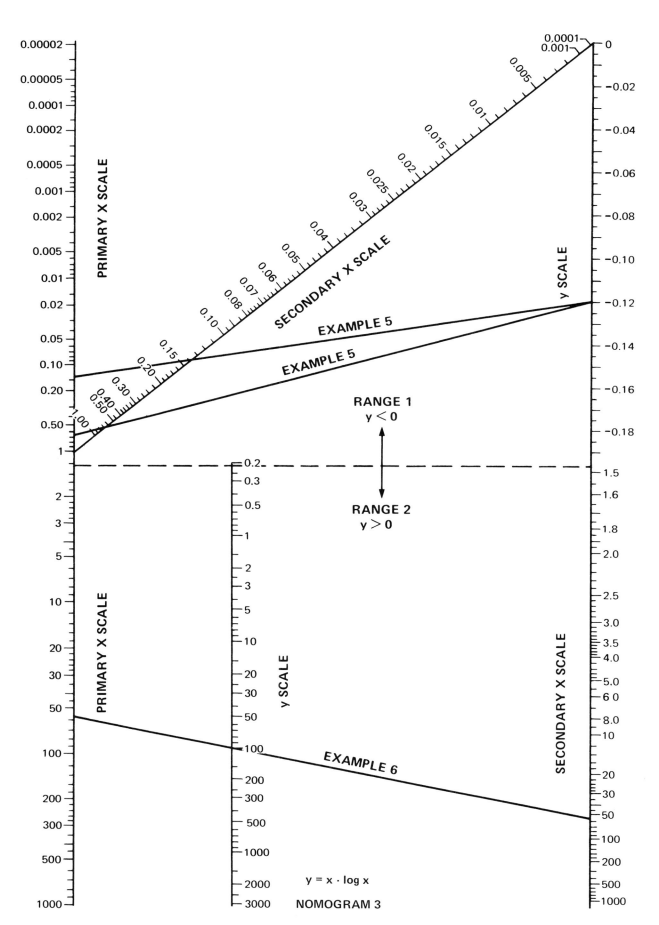

$y = x \cdot \log x$

NOMOGRAM 3

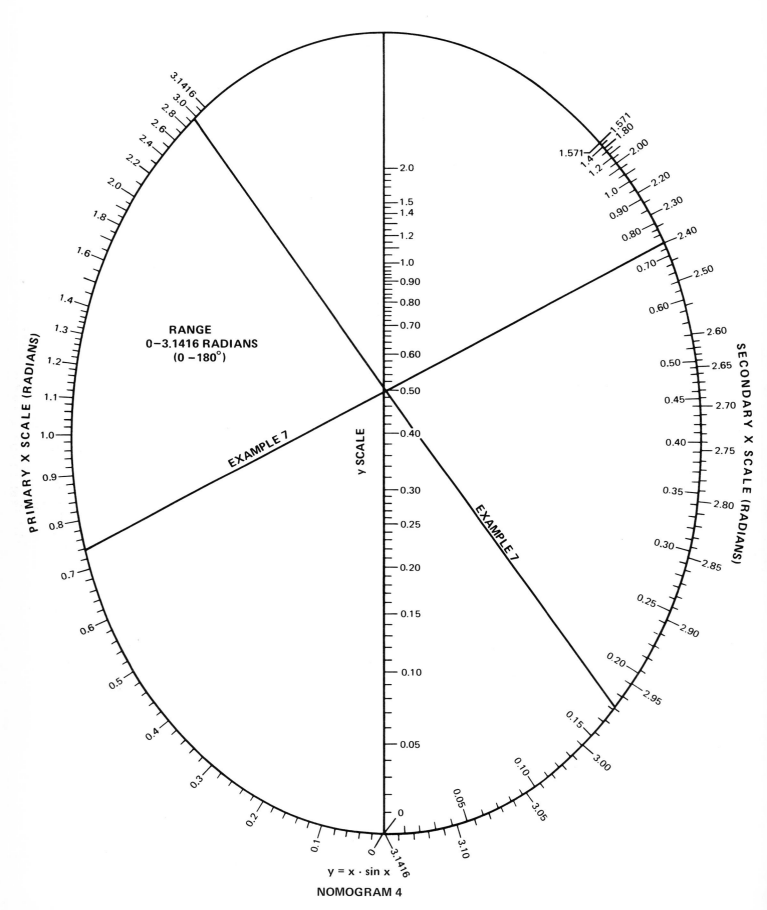

PRIMARY X SCALE (RADIANS)

SECONDARY X SCALE (RADIANS)

y SCALE

RANGE
0–3.1416 RADIANS
(0 –180°)

EXAMPLE 7

EXAMPLE 7

$y = x \cdot \sin x$

NOMOGRAM 4

144

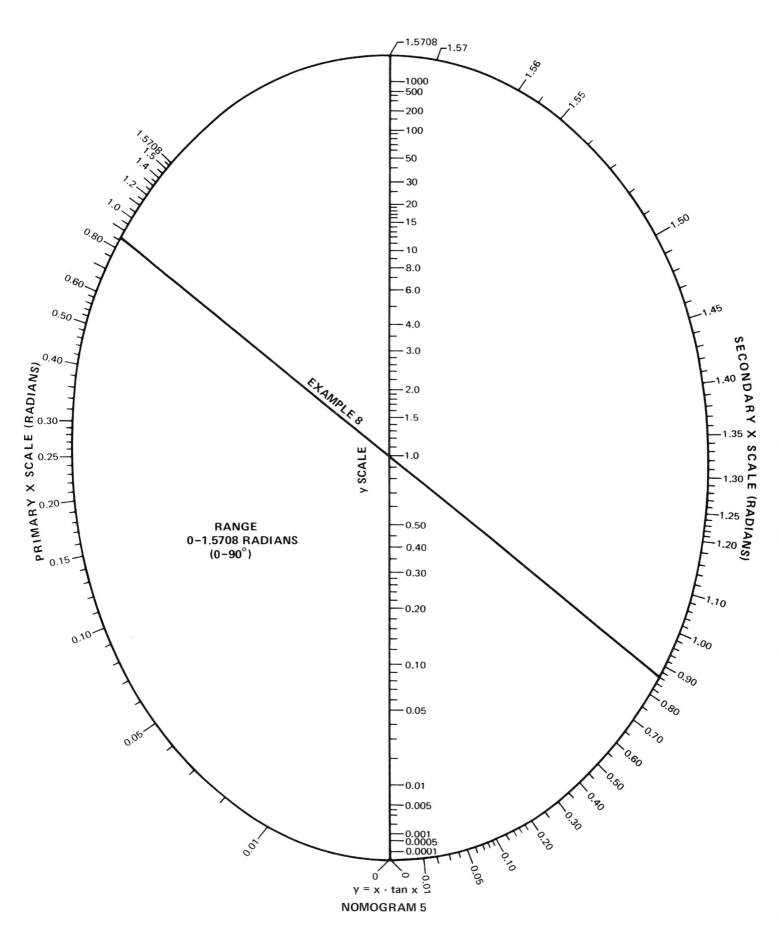

PRIMARY X SCALE (RADIANS)

SECONDARY X SCALE (RADIANS)

y SCALE

EXAMPLE 8

RANGE
0–1.5708 RADIANS
(0–90°)

$y = x \cdot \tan x$

NOMOGRAM 5

145

Nomogram for quadratic equations

Introduction.

Frequently in engineering calculations it is necessary to solve a quadratic equation. The following nomograms will expedite that process.

The equation fundamental to these nomograms is:

$$x = \frac{-b \pm \sqrt{b^2 - 4ac}}{2a}$$

for the general equation:

$$ax^2 \pm bx \pm c = 0$$

For use in these nomograms, the general equation has been cleared and simplified to:

$$x^2 \pm ax \pm b = 0 \qquad (1)$$

The nomograms in this article provide real roots and do not consider imaginary values.

Use of nomogram.

To use the nomograms, a value of Z is determined from Nomogram 1 and this value is carried over to the nomogram indicated. Inserting the Z value twice (as indicated by the signs in the equation) gives two answers, which are the roots of the equation.

Cases of quadratic equations.

The following four cases of quadratic equations are solved in this article:

Case 1.
$x^2 + ax + b = 0$
If $Z < 1$—Use Nomograms 5 and 6.
If $Z > 1$—x is imaginary

Case 2.
$x^2 - ax + b = 0$
If $Z < 1$—Use Nomograms 5 and 6
If $Z > 1$—x is imaginary

Case 3.
$x^2 + ax - b = 0$
If $Z < 1$—Use Nomograms 2 and 3
If $Z > 1$—Use Nomograms 3 and 4

Case 4.
$x^2 - ax - b = 0$
If $Z < 1$—Use Nomograms 2 and 3
If $Z > 1$—Use Nomograms 3 and 4

Nomogram 1 procedure.

Construct a line from a on the a scale to b on the b scale intersecting Z on the Z scale. Adjust the decimal point if necessary and proceed to the nomogram indicated.

Nomogram 2 procedure.

Construct a line from a on the a scale to Z on the Z scale and continue this line until it intersects x on the x scale.
For $x^2 + ax - b = 0$, read x as plus
For $x^2 - ax - b = 0$, read x as minus

Nomogram 3 procedure.

Construct a line from a on the a scale to Z on the Z scale (on the Z scale the value of Z is entered once on the minus side of the scale and once on the plus side of the scale) intersecting the x scale twice.

For $x^2 + ax - b = 0$, carry the sign of the Z scale used
For $x^2 - ax - b = 0$, reverse the sign of the Z scale used

Nomogram 4 procedure.

Construct a line from a on the a scale to Z on the Z scale (on the Z scale the value of Z is entered once on the minus side of the scale and once on the plus side of the scale) intersecting the x scale twice.

For $x^2 + ax - b = 0$, carry the sign of the Z scale used
For $x^2 - ax - b = 0$, reverse the sign of the Z scale used

Nomogram 5 procedure.

Construct a line from a on the a scale to Z on the Z scale intersecting the x scale.

For $x^2 + ax + b = 0$, read x as minus
For $x^2 - ax + b = 0$, read x as plus

Nomogram 6 procedure.

Construct a line from a on the a scale to Z on the Z scale intersecting x on the x scale (read x on the same side of the scale as Z is entered). For values of Z below 0.001, extract the decimal to enter the Z scale and adjust the decimal as indicated. For such values use the 0.001 to 0.010 section of the Z scale.

For $x^2 - ax + b = 0$, read x as plus
For $x^2 + ax + b = 0$, read x as minus

Example 1:

Solve the following quadratic equation:

$$x^2 + x - 6 = 0$$

Solution 1:

From inspection of this equation (compared to Eq. 1), the following values are obtained:

$$a = 1 \qquad b = 6$$

Proceeding to Nomogram 1, we construct a line from 1 on the a scale to 6 on the b scale intersecting the Z scale at 24. We next examine the equation and observe that it falls into a Case 3 quadratic equation. We also observe that $Z > 1$, thus we proceed to Nomograms 3 and 4 (Nomogram 4 has larger x and Z values than Nomogram 3 but since in this problem we are talking about low values, it isn't necessary to go to Nomogram 4). Using Nomogram 3, we construct a line from $a = 1$ on the a scale to $Z = 24$ on the Z scale (on the Z scale the value of Z is entered once on the minus side of the scale and once on the plus side of the scale) intersecting the x scale at $x = 2$ and $x = 3$ respectively. Thus the real roots are:

$$x = +2$$
$$x = -3$$

Example 2:

Solve the following quadratic equation:

$$5x^2 - 7500x + 2,500,000 = 0$$

Solution 2:

Reduction of the equation gives us:

$$x^2 - 1500x + 500,000 = 0$$

From inspection of this equation (compared to Eq. 1), the following values are obtained:

$$a = 1.5 \times 10^3 \qquad b = 5 \times 10^5$$

Proceeding to Nomogram 1, we construct a line from 1.5 on the a scale to 5 on the b scale intersecting the Z scale at 8.9. Adjusting the decimal $(10^{m-2n}) = (10^{5-(2)(3)})$ gives $Z = 0.89$. We next examine the equation and observe that it falls into a Case 2 quadratic equation. We also observe that $Z < 1$, thus we proceed to Nomograms 5 and 6. Using Nomogram 5, construct a line from $a = 1.5$ on the a scale to $Z = 0.89$ on the Z scale intersecting the x scale at $x = 1.0$. Adjusting the decimal gives us $x = 1000$. As indicated in the Nomogram 5 procedure, since a is minus, x is a positive root whose value is $x = +1000$. Using Nomogram 6, construct a line from $a = 1.5$ on the a scale to $Z = 0.89$ on the Z scale intersecting the x scale at $x = 0.5$. Adjusting the decimal gives us $x = 500$. As indicated in the Nomogram 6 procedure, since a is minus, x has a positive root whose value is $x = +500$. Thus the real roots are:

$$x = +1000$$
$$x = +500$$

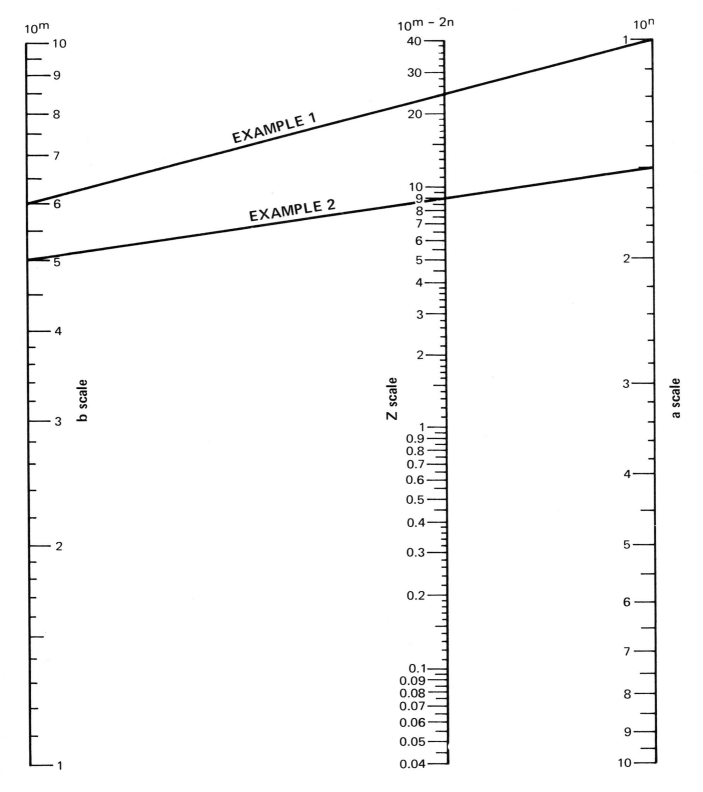

QUADRATIC NOMOGRAM 1

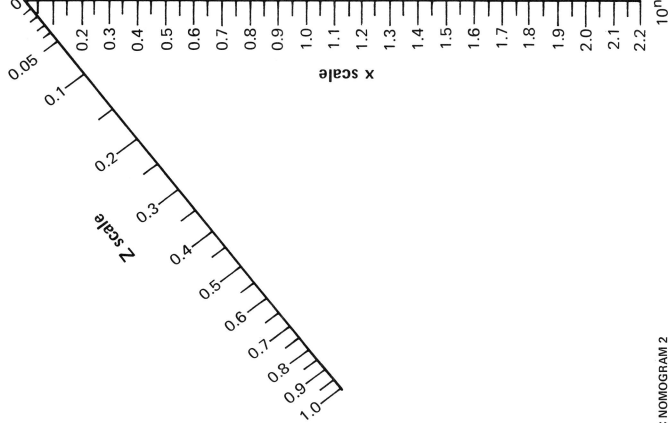

x scale

Z scale

a scale

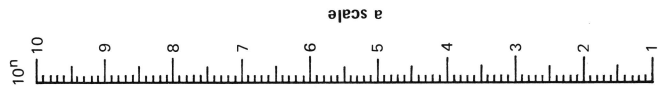

QUADRATIC NOMOGRAM 2

149

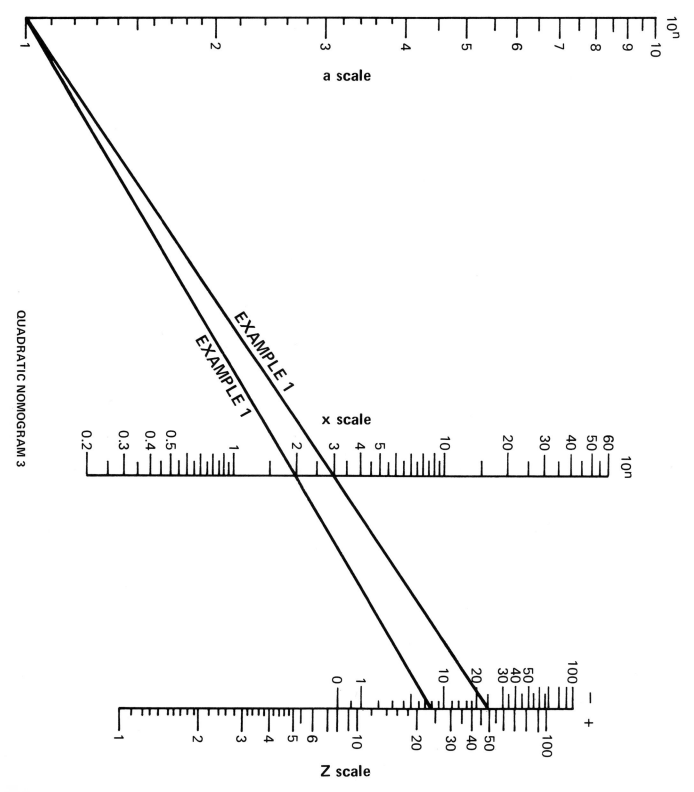

QUADRATIC NOMOGRAM 3

a scale

x scale

Z scale

EXAMPLE 1

EXAMPLE 1

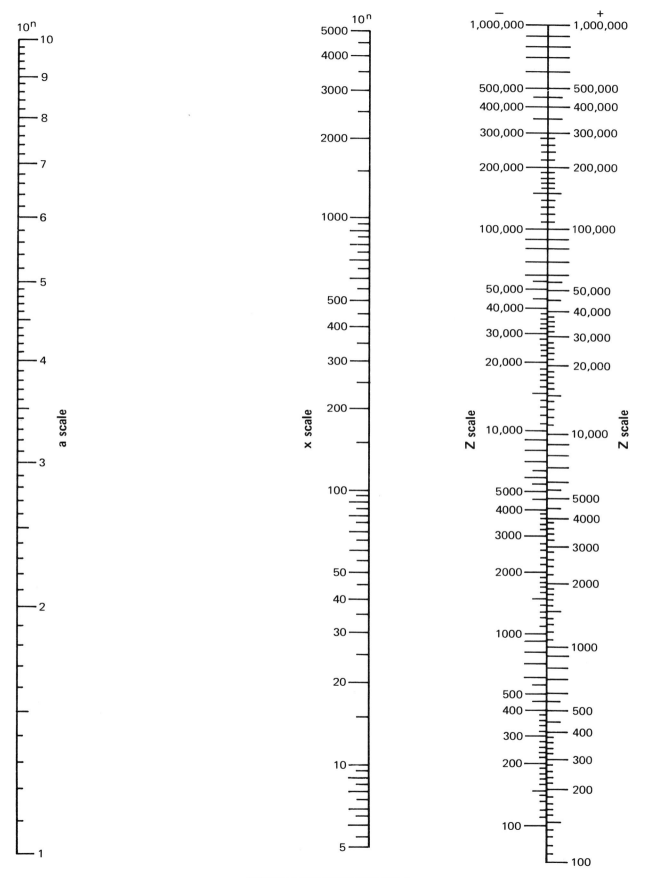

10^n

a scale

10^n

x scale

− +

Z scale   Z scale

QUADRATIC NOMOGRAM 4

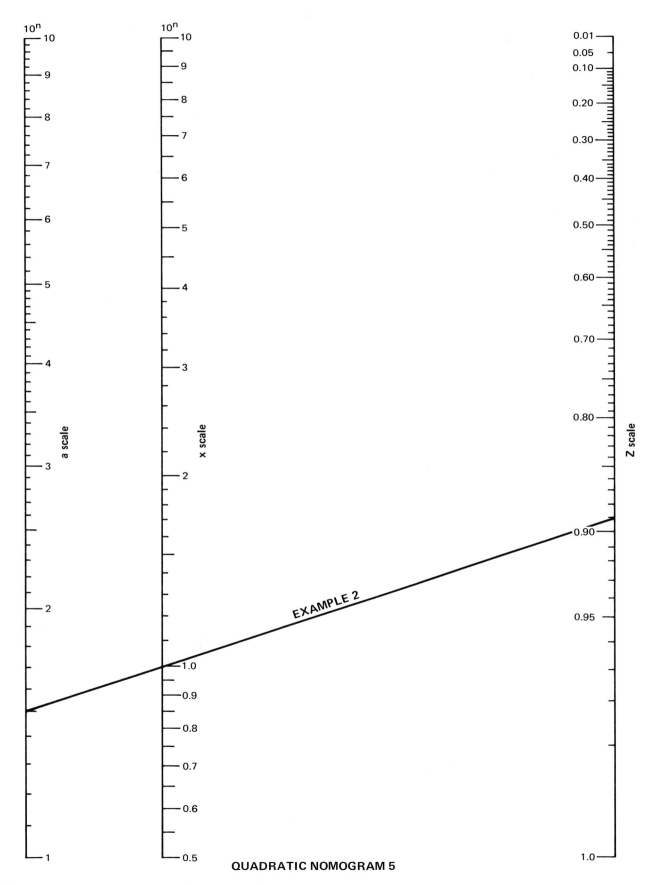

10^n

10^n

a scale

x scale

Z scale

EXAMPLE 2

QUADRATIC NOMOGRAM 5

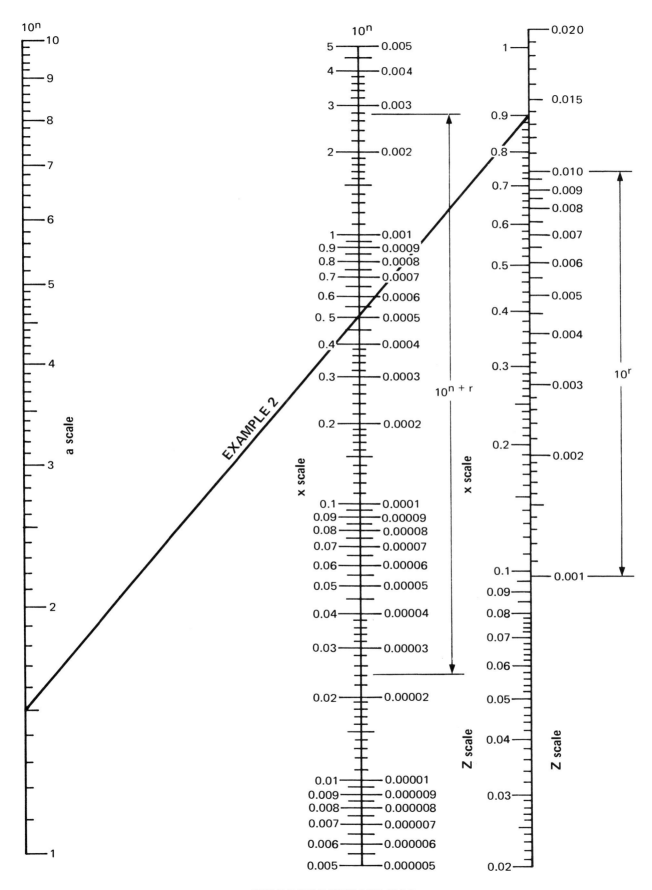

QUADRATIC NOMOGRAM 6

Geometrical curves and their equations

Introduction.

Frequently in engineering calculations, the shape and equation of a geometrical curve is required. The following curves represent the ones most commonly used.

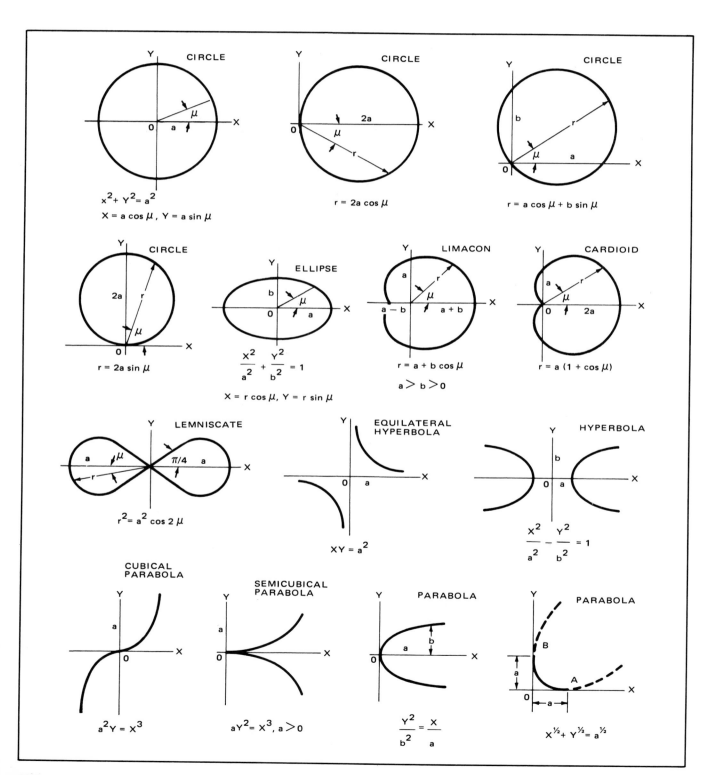

CIRCLE

$$x^2 + Y^2 = a^2$$
$$X = a \cos \mu, \ Y = a \sin \mu$$

CIRCLE

$$r = 2a \cos \mu$$

CIRCLE

$$r = a \cos \mu + b \sin \mu$$

CIRCLE

$$r = 2a \sin \mu$$

ELLIPSE

$$\frac{X^2}{a^2} + \frac{Y^2}{b^2} = 1$$
$$X = r \cos \mu, \ Y = r \sin \mu$$

LIMACON

$$r = a + b \cos \mu$$
$$a > b > 0$$

CARDIOID

$$r = a \ (1 + \cos \mu)$$

LEMNISCATE

$$r^2 = a^2 \cos 2 \mu$$

EQUILATERAL HYPERBOLA

$$XY = a^2$$

HYPERBOLA

$$\frac{X^2}{a^2} - \frac{Y^2}{b^2} = 1$$

CUBICAL PARABOLA

$$a^2 Y = X^3$$

SEMICUBICAL PARABOLA

$$aY^2 = X^3, \ a > 0$$

PARABOLA

$$\frac{Y^2}{b^2} = \frac{X}{a}$$

PARABOLA

$$X^{1/2} + Y^{1/2} = a^{1/2}$$

Example 1:

Describe the shape of a lemniscate and its equation.

Solution 1:

Looking through the chart we locate the lemniscate. It has the shape as shown and its equation is:

$$r^2 = a^2 \cos 2\mu$$

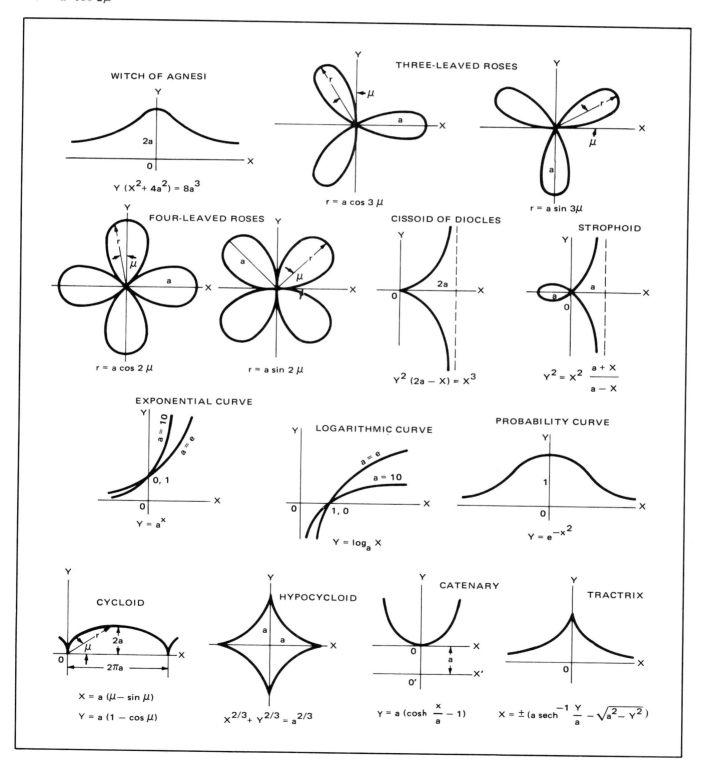

Numerical integration for area under a curve

Introduction.
A very useful equation in mathematics is Simpson's Rule, one form of which can be written (Fig. 1):

$$\int_{x_0}^{x_2} ydx = \frac{m}{3}[y_0 + 4y_1 + y_2] \tag{1}$$

The integral on the left denotes the area under the curve from x_0 to x_2 and the y's represent the corresponding ordinates to the curve. Distance m represents the horizontal spacing of the ordinates.

Nomenclature:

y_0 = Area of base
y_2 = Area of top surface
y_1 = Area of cross-section halfway between.

  y_1 is usually not equal to $\frac{1}{2}(y_0 + y_2)$

m = Horizontal distance from y_0 to y_1 and from y_1 to y_2
V = Volume
A = Area

Example 1:
Given a parabola (dimensions given in Fig. 2), determine the area under the curve between values of x of 2.0 and 4.0.

Solution 1:
The ordinates are found as follows:

at x = 2.0, $y_2 = x^2$
      $y_2 = 2.0^2$
      $y_2 = 4.0$
at x = 3.0, $y_3 = x^2$
      $y_3 = 3.0^2$
      $y_3 = 9.0$
at x = 4.0, $y_4 = x^2$
      $y_4 = 4.0^2$
      $y_4 = 16.0$
  m = 1.0

Here y_2, y_3 and y_4 must be substituted for y_0, y_1 and y_2 respectively in Eq. 1. Thus:

$$A = \int_{2.0}^{4.0} ydx = \frac{m}{3}[y_2 + 4y_3 + y_4]$$
$$A = \frac{1.0}{3}[4.0 + (4)(9.0) + 16.0]$$
$$A = \frac{56}{3}$$
$$A = 18.6 \text{ ft}^2$$

Example 2:
Given a frustum of a pyramid (dimensions given in Fig. 3), determine its volume.

Solution 2:
The computations are as follows:

$y_0 = (10.0)(8.0)$
$y_0 = 80.0$
$y_2 = (6.0)(4.0)$
$y_2 = 24.0$
$y_1 = (6.0)(8.0)$
$y_1 = 48.0$

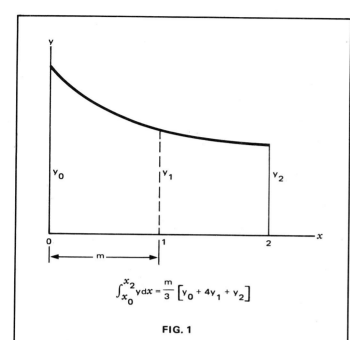

$$\int_{x_0}^{x_2} ydx = \frac{m}{3}\left[y_0 + 4y_1 + y_2\right]$$

FIG. 1

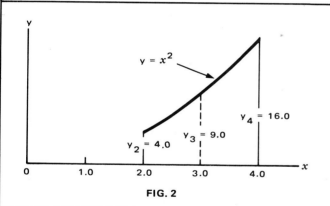

FIG. 2

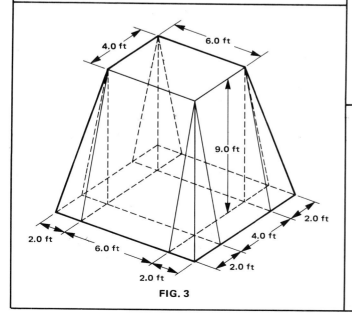

FIG. 3

$$m = \frac{1}{2}(9.0)$$
$$m = 4.5$$

Substituting into Eq. 1:

$$V = \int_0^{9.0} ydx = \frac{m}{3}\left[y_0 + 4y_1 + y_2\right]$$
$$V = \frac{4.5}{3}\left[80.0 + (4)(48.0) + 24.0\right]$$
$$V = 444.0 \text{ ft}^3$$

This result could have been easily obtained by integration without the use of Simpson's Rule, but there are many cases where the curve has a complicated form and integration is difficult if not impossible, and it is here that Simpson's rule is of real utility. For such curves, more terms in the equation are needed on the right and the equation assumes the following forms:

For Fig. 4:

$$\int_{x_0}^{x_4} ydx = \frac{m}{3}\left[y_0 + 4y_1 + 2y_2 + 4y_3 + y_4\right]$$

For Fig. 5:

$$\int_{x_0}^{x_6} ydx = \frac{m}{3}\left[y_0 + 4y_1 + 2y_2 + 4y_3 + 2y_4 + 4y_5 + y_6\right]$$

If more terms are needed, it is merely necessary to follow the above routine when writing a new equation. The result may only be an approximation, but it can be a very good one if a sufficient number of terms are taken. Note that the horizontal interval is divided into an even number of spaces of length m.

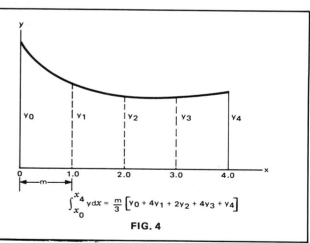

$$\int_{x_0}^{x_4} ydx = \frac{m}{3}\left[y_0 + 4y_1 + 2y_2 + 4y_3 + y_4\right]$$

FIG. 4

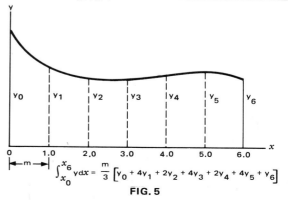

$$\int_{x_0}^{x_6} ydx = \frac{m}{3}\left[y_0 + 4y_1 + 2y_2 + 4y_3 + 2y_4 + 4y_5 + y_6\right]$$

FIG. 5

Differential and integral calculus equations

Introduction.
The following article contains the differential and integral equations most commonly used in the field of mathematics. In this article, u and v are functions of x, C and n are constants and e = 2.71828.

General equations.

1. $\frac{dC}{dx} = 0$ (C is a constant)

2. $\frac{dx}{dx} = 1$

3. $\frac{d}{dx} = (u + v + w + \ldots) = \frac{du}{dx} + \frac{dv}{dx} + \frac{dw}{dx} + \ldots$

4. $\frac{d}{dx}(Cu) = C\frac{du}{dx}$

5. $\frac{dy}{dx} = \frac{dy}{du} \cdot \frac{du}{dx}$

   $= \frac{dy}{dx} = \frac{1}{\frac{dy}{dx}}$

6. $\frac{d}{dx}\log_a u = \frac{1}{u} \cdot \log_a e \cdot \frac{du}{dx}$ (The base a is positive and not equal to 1)

7. $\frac{d}{dx}\ln u = \frac{1}{u}\frac{du}{dx}$ (ln u = $\log_e u$)

8. $\frac{d}{dx}(u^n) = nu^{n-1}\frac{du}{dx}$

9. $\frac{d}{dx}(uv) = v\frac{du}{dx} + u\frac{dv}{dx}$

10. $\frac{d}{dx}\left(\frac{u}{v}\right) = \frac{v\frac{du}{dx} - u\frac{dv}{dx}}{v^2}$

11. $\frac{d}{dx}(uvw) = uv\frac{d}{dx}(w) + uw\frac{d}{dx}(v) + vw\frac{d}{dx}(u)$

12. $\frac{d}{dx}\left(\frac{u}{C}\right) = \frac{1}{C} \cdot \frac{d}{dx}(u)$

13. $\frac{d}{dx}\left(\frac{C}{u}\right) = C\frac{d}{dx}\left(\frac{1}{u}\right) = -\frac{C}{u^2} \cdot \frac{d}{dx}(u)$

14. $\frac{d}{dx}(x^n) = nx^{n-1}$

15. $d(u^m) = mu^{m-1}du$

16. $d\sqrt{u} = \frac{du}{2\sqrt{u}}$

17. $d\left(\frac{1}{u}\right) = -\frac{du}{u^2}$

18. $d(e^u) = e^u du$

19. $d(a^u) = (\ln a)a^u du$

20. $d\ln u = \frac{du}{u}$

21. $d\log_{10} u = (\log_{10} e)\frac{du}{u} = (0.4343\ldots)\frac{du}{u}$

Trigonometric functions.

22. $\dfrac{d}{dx} \sin u = \cos u \dfrac{du}{dx}$ $d \sin u = \cos u \, du$

23. $\dfrac{d}{dx} \cos u = -\sin u \dfrac{du}{dx}$ $d \cos \bar{u} = -\sin u \, du$

24. $\dfrac{d}{dx} \tan u = \sec^2 u \dfrac{du}{dx}$ $d \tan u = \sec^2 u \, du$

25. $\dfrac{d}{dx} \cot u = -\csc^2 u \dfrac{du}{dx}$ $d \cot u = -\csc^2 u \, du$

26. $\dfrac{d}{dx} \sec u = \sec u \tan u \dfrac{du}{dx}$ $d \sec u = \sec u \tan u \, du$

27. $\dfrac{d}{dx} \csc u = -\csc u \cot u \dfrac{du}{dx}$ $d \csc u = -\csc u \cot u \, du$

Inverse trigonometric functions.

28. $\dfrac{d}{dx} \sin^{-1} u = \dfrac{1}{\sqrt{1-u^2}} \dfrac{du}{dx}$ $d \sin^{-1} u = \dfrac{du}{\sqrt{1-u^2}}$

29. $\dfrac{d}{dx} \cos^{-1} u = -\dfrac{1}{\sqrt{1-u^2}} \dfrac{du}{dx}$ $d \cos^{-1} u = -\dfrac{du}{\sqrt{1-u^2}}$

30. $\dfrac{d}{dx} \tan^{-1} u = \dfrac{1}{1+u^2} \dfrac{du}{dx}$ $d \tan^{-1} u = \dfrac{du}{1+u^2}$

31. $\dfrac{d}{dx} \cot^{-1} u = -\dfrac{1}{1+u^2} \dfrac{du}{dx}$ $d \cot^{-1} u = -\dfrac{du}{1+u^2}$

32. $\dfrac{d}{dx} \sec^{-1} u = \dfrac{1}{u\sqrt{u^2-1}} \dfrac{du}{dx}$ $d \sec^{-1} u = \dfrac{du}{u\sqrt{u^2-1}}$

33. $\dfrac{d}{dx} \csc^{-1} u = -\dfrac{1}{u\sqrt{u^2-1}} \dfrac{du}{dx}$ $d \csc^{-1} u = -\dfrac{du}{u\sqrt{u^2-1}}$

Exponential functions.

34. $\dfrac{d}{dx} a^u = a^u \ln a \dfrac{du}{dx}$ $d\, a^u = a^u \ln a \, du$

35. $\dfrac{d}{dx} e^u = e^u \dfrac{du}{dx}$ $d\, e^u = e^u \, du$

36. $\dfrac{d}{dx} u^v = v u^{v-1} \dfrac{du}{dx} + u^v \ln u \dfrac{dv}{dx}, \, du^v = v u^{v-1} \, du + u^v \ln u \, dv$

Miscellaneous expressions.

37. $d \ln \sin u = \cot u \, du$
38. $d \ln \cos u = -\tan u \, du$

39. $d \ln \tan u = \dfrac{2\, du}{\sin 2u}$

40. $d \ln \cot u = -\dfrac{2\, du}{\sin 2u}$

41. $d \sinh u = \cosh u \, du$
42. $d \cosh u = \sinh u \, du$
43. $d \tanh u = \operatorname{sech}^2 u \, du$
44. $d \operatorname{csch} u = -\operatorname{csch} u \coth u \, du$
45. $d \operatorname{sech} u = -\operatorname{sech} u \tanh u \, du$
46. $d \coth u = -\operatorname{csch}^2 u \, du$

47. $d \sinh^{-1} u = \dfrac{du}{\sqrt{u^2+1}}$

48. $d \cosh^{-1} u = \dfrac{du}{\sqrt{u^2-1}}$

49. $d \tanh^{-1} u = \dfrac{du}{1-u^2}$

50. $d \operatorname{csch}^{-1} u = -\dfrac{du}{u\sqrt{u^2+1}}$

51. $d \operatorname{sech}^{-1} u = -\dfrac{du}{u\sqrt{1-u^2}}$

52. $d \coth^{-1} u = \dfrac{du}{1-u^2}$

53. $d(u^v) = (u^{v-1})(u \ln u \, dv + v \, du)$

First differential coefficient of functions of one variable.
If equation on left is given, then equation on right represents its first differential coefficient of the function of one variable.

54. $y = ax^n$ $\dfrac{dy}{dx} = anx^{n-1}$
(Whatever be the value of n)

55. $y = a^x$ $\dfrac{dy}{dx} = (p)(a^x) = \log[(a)(a^x)]$

56. $y = \log x$ $\dfrac{dy}{dx} = \left[\left(\dfrac{1}{p}\right)\left(\dfrac{1}{x}\right)\right]$

57. $y = \sin x$ $\dfrac{dy}{dx} = \cos x$

58. $y = \cos x$ $\dfrac{dy}{dx} = -\sin x$

59. $y = e^x$ $\dfrac{dy}{dx} = e^x$

60. $y = \log_e x$ $\dfrac{dy}{dx} = \dfrac{1}{x}$

61. $y = \sin mx$ $\dfrac{dy}{dx} = m \cos mx$

62. $y = \cos mx$ $\dfrac{dy}{dx} = -m \sin mx$

63. $y = x^n \pm b$ $\dfrac{dy}{dx} = nx^{n-1}$

64. $y = a^x \pm b$ $\dfrac{dy}{dx} = \log(a)(a^x)$

65. $y = e^x \pm b$ $\dfrac{dy}{dx} = e^x$

66. $y = a + x^{-q}$ $\dfrac{dy}{dx} = -qx^{-(q+1)}$

67. $y = \log x \pm b$ $\dfrac{dy}{dx} = \left(\dfrac{1}{\log a}\right)\left(\dfrac{1}{x}\right)$

68. $y = \log_e \pm b$ $\dfrac{dy}{dx} = \dfrac{1}{x}$

69. $y = \sin x \pm b$ $\dfrac{dy}{dx} = \cos x$

70. $y = \cos x \pm b$ $\dfrac{dy}{dx} = -\sin x$

71. $y = cx^n$ $\quad\dfrac{dy}{dx} = ncx^{n-1}$

72. $y = ca^x$ $\quad\dfrac{dy}{dx} = c \log a(a)$

73. $y = (c)(e^x)$ $\quad\dfrac{dy}{dx} = (c)(e^x)$

74. $y = c \log x$ $\quad\dfrac{dy}{dx} = \left(\dfrac{c}{\log a}\right)\left(\dfrac{1}{x}\right)$

75. $y = c \sin x$ $\quad\dfrac{dy}{dx} = c \cos x$

76. $y = c \cos x$ $\quad\dfrac{dy}{dx} = -c \sin x$

Inverse functions.
If equation on left is given, then equation on right represents its inverse function.

77. $y = \sin^{-1}x$ $\quad\dfrac{dy}{dx} = \dfrac{1}{\sqrt{1-x^2}}$

78. $y = \cos^{-1}x$ $\quad\dfrac{dy}{dx} = -\dfrac{1}{\sqrt{1-x^2}}$

79. $y = \tan^{-1}x$ $\quad\dfrac{dy}{dx} = \dfrac{1}{1+x^2}$

80. $y = \cot^{-1}x$ $\quad\dfrac{dy}{dx} = -\dfrac{1}{1+x^2}$

81. $y = \sec^{-1}x$ $\quad\dfrac{dy}{dx} = \dfrac{1}{x\sqrt{x^2-1}}$

82. $y = \csc^{-1}x$ $\quad\dfrac{dy}{dx} = -\dfrac{1}{x\sqrt{x^2-1}}$

83. $y = \text{vers}^{-1}x$ $\quad\dfrac{dy}{dx} = \dfrac{1}{\sqrt{2x-x^2}}$

84. $y = \text{chd}^{-1}x$ $\quad\dfrac{dy}{dx} = \dfrac{2}{\sqrt{4-x^2}}$

85. $y = \sin^{-1}x$ $\quad\dfrac{dy}{dx} = \dfrac{r}{\sqrt{r^2-x^2}}$ (R = radius)

86. $y = \cos^{-1}x$ $\quad\dfrac{dy}{dx} = -\dfrac{r}{\sqrt{r^2-x^2}}$

87. $y = \text{vers}^{-1}x$ $\quad\dfrac{dy}{dx} = \dfrac{r}{\sqrt{2rx-x^2}}$

88. $y = \tan^{-1}x$ $\quad\dfrac{dy}{dx} = \dfrac{r^2}{r^2+x^2}$

89. $y = \cot^{-1}x$ $\quad\dfrac{dy}{dx} = -\dfrac{r^2}{r^2+x^2}$

90. $y = \sec^{-1}x$ $\quad\dfrac{dy}{dx} = \dfrac{r^2}{x\sqrt{x^2-r^2}}$

91. $y = \csc^{-1}x$ $\quad\dfrac{dy}{dx} = -\dfrac{r^2}{x\sqrt{x^2-r^2}}$

92. $y = \text{chd}^{-1}x$ $\quad\dfrac{dy}{dx} = \dfrac{2r}{\sqrt{4r^2-x^2}}$

93. $\sin^{-1}x = \cos^{-1}\sqrt{1-x^2} = \tan^{-1}\dfrac{x}{\sqrt{1-x^2}}$

$\qquad = \cot^{-1}\dfrac{\sqrt{1-x^2}}{x} = \sec^{-1}\dfrac{1}{\sqrt{1-x^2}} = \csc^{-1}\dfrac{1}{x}$

94. $\cos^{-1}x = \sin^{-1}\sqrt{1-x^2} = \tan^{-1}\dfrac{\sqrt{1-x^2}}{x}$

$\qquad = \cot^{-1}\dfrac{x}{\sqrt{1-x^2}} = \sec^{-1}\dfrac{1}{x} = \csc^{-1}\dfrac{1}{\sqrt{1-x^2}}$

95. $\tan^{-1}x = \sin^{-1}\dfrac{x}{\sqrt{1+x^2}} = \cos^{-1}\dfrac{1}{\sqrt{1+x^2}} = \cot^{-1}\dfrac{1}{x}$

$\qquad = \sec^{-1}\sqrt{1+x^2} = \csc^{-1}\dfrac{\sqrt{1+x^2}}{x}$

96. $\cot^{-1}x = \tan^{-1}\dfrac{1}{x}$, $\sec^{-1}x = \cos^{-1}\dfrac{1}{x}$, $\csc^{-1}x = \sin^{-1}\dfrac{1}{x}$

97. $y = \sqrt{\csc x}$ $\quad\dfrac{dy}{dx} = -\dfrac{\cos x}{2\sin^{3/2}x}$

98. $y = \sin^{-1}\dfrac{1-x^2}{1+x^2}$ $\quad\dfrac{dy}{dx} = -\dfrac{2}{1+x^2}$

99. $y = \cot^{-1}\sqrt{\dfrac{1-x}{x}}$ $\quad\dfrac{dy}{dx} = \dfrac{1}{2\sqrt{x-x^2}}$

100. $y = \log\left(\dfrac{1+\sqrt{-1}\,\tan x}{1-\sqrt{-1}\,\tan x}\right)^n$ $\quad\dfrac{dy}{dx} = 2n\sqrt{-1}$

Elementary forms of the integral equations

101. $\displaystyle\int du = u + C$

102. $\displaystyle\int(du + dv - dw) = \int du + \int dv - \int dw$

103. $\displaystyle\int a\, du = a\int du = au + C$

104. $\displaystyle\int u\, dv = uv - \int v\, du$

105. $\displaystyle\int(u + v)\, dx = \int u\, dx + \int v\, dx$

106. $\displaystyle\int f(x)\, dx = \int f[F(y)]\, F'(y)\, dy$, $x = F(y)$

107. $\displaystyle\int dy \int f(x,y)\, dx = \int dx \int f(x,y)\, dy$

Fundamental integrals.

108. $\displaystyle\int\dfrac{d}{dx}[f(x)]dx = f(x) + C$

109. $\displaystyle\int a^u\, du = \dfrac{a^u}{\ln a} + C$, if $a > 0$

110. $\displaystyle\int au\, dx = a\int u\, dx$, where a is any constant

111. $\displaystyle\int u^n\, du = \dfrac{u^{n+1}}{n+1} + C$, if $n \neq -1$

112. $\displaystyle\int u^{-1}\, du = \int\dfrac{du}{u} = \ln u + C$

113. $\displaystyle\int\dfrac{du}{u} = \ln u + C$, if $u > 0$

114. $\int \log x \, dx = x \log x - x + C$

115. $\int e^u \, du = e^u + C$

116. $\int \sin u \, du = -\cos u + C$

117. $\int \cos u \, du = \sin u + C$

118. $\int \tan u \, du = \ln \sec u + C$

119. $\int \cot u \, du = \ln \sin u + C$

120. $\int \sec u \, du = \ln (\sec u + \tan u) + C$

121. $\int \csc u \, du = \ln (\csc u - \cot u) + C$

122. $\int \sec^2 u \, du = \tan u + C$

123. $\int \csc^2 u \, du = -\cot u + C$

124. $\int \sec u \tan u \, du = \sec u + C$

125. $\int \csc u \cot u \, du = -\csc u + C$

126. $\int u \, dv = uv - \int v \, du$

Expressions involving $\sqrt{a^2 - x^2}$ and $\dfrac{du}{\sqrt{a^2 + u^2}}$.

127. $\int \dfrac{du}{\sqrt{a^2 - u^2}} = \sin^{-1} \dfrac{u}{a} + C$, if $u^2 < a^2$

128. $\int \dfrac{du}{\sqrt{1 - u^2}} = \sin^{-1} u + C$

129. $\int \dfrac{du}{a^2 + u^2} = \dfrac{1}{a} \tan^{-1} \dfrac{u}{a} + C$

130. $\int \dfrac{du}{1 + u^2} = \tan^{-1} u + C$

131. $\int \dfrac{du}{u \sqrt{u^2 - a^2}} = \dfrac{1}{a} \sec^{-1} \dfrac{u}{a} + C$, if $u^2 > a^2$

132. $\int \dfrac{du}{u \sqrt{u^2 - 1}} = \sec^{-1} u + C$

133. $\int \dfrac{du}{u^2 - a^2} = \dfrac{1}{2a} \ln \dfrac{u - a}{u + a} + C$, if $u^2 > a^2$

134. $\int \dfrac{du}{a^2 - u^2} = \dfrac{1}{2a} \ln \dfrac{a + u}{a - u} + C$, if $u^2 < a^2$

135. $\int \dfrac{du}{\sqrt{u^2 + a^2}} = \ln (u + \sqrt{u^2 + a^2}) + C$

136. $\int \dfrac{du}{\sqrt{u^2 - a^2}} = \ln (u + \sqrt{u^2 - a^2}) + C$, if $u^2 > a^2$

137. $\int \sqrt{a^2 - u^2} \, du = \dfrac{1}{2} u \sqrt{a^2 - u^2} + \dfrac{1}{2} a^2 \sin^{-1} \dfrac{u}{a} + C$

138. $\int \dfrac{dx}{\sqrt{a^2 - x^2}} = \sin^{-1} \dfrac{x}{a} + C$

139. $\int \dfrac{x \, dx}{\sqrt{a^2 - x^2}} = -\sqrt{a^2 - x^2} + C$

140. $\int \dfrac{x^2 \, dx}{\sqrt{a^2 - x^2}} = -\dfrac{x}{2} \sqrt{a^2 - x^2} + \dfrac{a^2}{2} \sin^{-1} \dfrac{x}{a} + C$

141. $\int \dfrac{x^3 \, dx}{\sqrt{a^2 - x^2}} = -\dfrac{1}{3} (x^2 + 2a^2) \sqrt{a^2 - x^2} + C$

142. $\int \dfrac{dx}{x \sqrt{a^2 - x^2}} = -\dfrac{1}{a} \ln \dfrac{a + \sqrt{a^2 - x^2}}{x} + C$

143. $\int \dfrac{dx}{x^2 \sqrt{a^2 - x^2}} = -\dfrac{\sqrt{a^2 - x^2}}{a^2 x} + C$

144. $\int \dfrac{dx}{x^3 \sqrt{a^2 - x^2}} = -\dfrac{\sqrt{a^2 - x^2}}{2a^2 x^2} - \dfrac{1}{2a^3} \ln \dfrac{a + \sqrt{a^2 - x^2}}{x} + C$

145. $\int \sqrt{a^2 - x^2} \, dx = \dfrac{x}{2} \sqrt{a^2 - x^2} + \dfrac{a^2}{2} \sin^{-1} \dfrac{x}{a} + C$

146. $\int x \sqrt{a^2 - x^2} \, dx = -\dfrac{1}{3} (a^2 - x^2)^{3/2} + C$

147. $\int x^2 \sqrt{a^2 - x^2} \, dx = -\dfrac{x}{4} (a^2 - x^2)^{3/2} + \dfrac{a^2}{8} (x \sqrt{a^2 - x^2} + a^2 \sin^{-1} \dfrac{x}{a}) + C$

148. $\int x^3 \sqrt{a^2 - x^2} \, dx = \left(-\dfrac{1}{5} x^2 - \dfrac{2}{15} a^2 \right) (a^2 - x^2)^{3/2} + C$

Expressions involving $\sqrt{x^2 \pm a^2}$ and $x \sqrt{x^2 - a^2}$.

149. $\int \sqrt{u^2 \pm a^2} \, du = \dfrac{1}{2} u \sqrt{u^2 \pm a^2} \pm \dfrac{1}{2} a^2 \ln (u + \sqrt{u^2 \pm a^2}) + C$

150. $\int \dfrac{du}{\sqrt{u^2 \pm a^2}} = \ln (u + \sqrt{u^2 \pm a^2}) + C$

151. $\int \dfrac{dx}{\sqrt{x^2 \pm a^2}} = \log (x + \sqrt{x^2 \pm a^2}) + C$

152. $\int \dfrac{x \, dx}{\sqrt{x^2 \pm a^2}} = \sqrt{x^2 \pm a^2} + C$

153. $\int \dfrac{x^2 \, dx}{\sqrt{x^2 \pm a^2}} = \dfrac{x}{2} \sqrt{x^2 \pm a^2} \mp \dfrac{a^2}{2} \ln (x + \sqrt{x^2 \pm a^2}) + C$

154. $\int \dfrac{x^3 \, dx}{\sqrt{x^2 \pm a^2}} = \dfrac{1}{3} (x^2 \mp 2 a^2) \sqrt{x^2 \pm a^2} + C$

155. $\int \dfrac{dx}{x \sqrt{x^2 + a^2}} = -\dfrac{1}{a} \ln \dfrac{a + \sqrt{x^2 + a^2}}{x} + C$

156. $\int \dfrac{dx}{x \sqrt{x^2 - a^2}} = \dfrac{1}{a} \sec^{-1} \dfrac{x}{a} + C$

157. $\int \dfrac{dx}{\sqrt{a^2 - x^2}} = \sin^{-1} \dfrac{x}{a} + C = -\cos^{-1} \dfrac{x}{a} + C$

158. $\int \dfrac{dx}{\sqrt{a^2 + x^2}} = \ln (x + \sqrt{a^2 + x^2}) + C = \sinh^{-1} \dfrac{x}{a} + C$

159. $\int \dfrac{dx}{\sqrt{x^2 - a^2}} = \ln (x + \sqrt{x^2 - a^2}) + C = \cosh^{-1} \dfrac{x}{a} + C$

161

160. $\int \sqrt{a^2 + x^2}\, dx = \dfrac{x}{2} \sqrt{a^2 + x^2} + \dfrac{a^2}{2} \ln (x + \sqrt{a^2 + x^2}) + C$

$\qquad = \dfrac{x}{2} \sqrt{a^2 + x^2} + \dfrac{a^2}{2} \sinh^{-1} \dfrac{x}{a} + C$

161. $\int \sqrt{a^2 - x^2}\, dx = \dfrac{x}{2} \sqrt{a^2 - x^2} + \dfrac{a^2}{2} \sin^{-1} \dfrac{x}{a} + C$

162. $\int \sqrt{x^2 - a^2}\, dx = \dfrac{x}{2} \sqrt{x^2 - a^2} - \dfrac{a^2}{2} \ln (x + \sqrt{x^2 - a^2}) + C$

$\qquad = \dfrac{x}{2} \sqrt{x^2 - a^2} - \dfrac{a^2}{2} \cosh^{-1} \dfrac{x}{a} + C$

163. $\int \dfrac{dx}{x^2 \sqrt{x^2 \pm a^2}} = \mp \dfrac{\sqrt{x^2 \pm a^2}}{a^2\, x} + C$

164. $\int \dfrac{dx}{x^3 \sqrt{x^2 + a^2}} = -\dfrac{\sqrt{x^2 + a^2}}{2\, a^2 x^2} + \dfrac{1}{2a^3} \ln \dfrac{a + \sqrt{x^2 + a^2}}{x} + C$

165. $\int \dfrac{dx}{x^3 \sqrt{x^2 - a^2}} = \dfrac{\sqrt{x^2 - a^2}}{2\, a^2 x^2} + \dfrac{1}{2a^3} \sec^{-1} \dfrac{x}{a} + C$

166. $\int \sqrt{x^2 \pm a^2}\, dx = \dfrac{x}{2} \sqrt{x^2 \pm a^2} \pm \dfrac{a^2}{2} \ln (x + \sqrt{x^2 \pm a^2} + C$

167. $\int x \sqrt{x^2 \pm a^2}\, dx = \dfrac{1}{3} (x^2 \pm a^2)^{3/2} + C$

168. $\int x^2 \sqrt{x^2 \pm a^2}\, dx = \dfrac{x}{4} (x^2 \pm a^2)^{3/2} \mp \dfrac{a^2}{8} x \sqrt{x^2 \pm a^2}$

$\qquad - \dfrac{a^4}{8} \ln (x + \sqrt{x^2 \pm a^2}) + C$

169. $\int x^3 \sqrt{x^2 \pm a^2}\, dx = \left(\dfrac{1}{5} x^2 \mp \dfrac{2}{15} a^2\right)(x^2 \pm a^2)^{3/2} + C$

Expressions involving $\sqrt{a + bx}$ and $\sqrt{\dfrac{a - x}{b + x}}$.

170. $\int \dfrac{x\, dx}{\sqrt{a + bx}} = -\dfrac{2\,(2a - bx)}{3\, b^2} \sqrt{a + bx} + C$

171. $\int \dfrac{x^2\, dx}{\sqrt{a + bx}} = \dfrac{2\,(8a^2 - 4abx + 3b^2x^2)}{15\, b^3} \sqrt{a + bx} + C$

172. $\int \sqrt{a + bx}\, dx = \dfrac{2}{3b} (\sqrt{a + bx})^3 + C$

173. $\int \dfrac{dx}{\sqrt{a + bx}} = \dfrac{2}{b} \sqrt{a + bx} + C$

174. $\int \dfrac{(m + nx)\, dx}{\sqrt{a + bx}} = \dfrac{2}{3b^2} (3mb - 2an + nbx) \sqrt{a + bx} + C$

175. $\int \sqrt{a + 2bx + cx^2}\, dx = \dfrac{b + cx}{2c} \sqrt{a + 2bx + cx^2}$

$\qquad + \dfrac{ac - b^2}{2c} \int \dfrac{dx}{\sqrt{a + 2bx + cx^2}} + C$

176. $\int x \sqrt{a + bx}\, dx = -\dfrac{2\,(2a - 3bx)}{15\, b^2} (a + bx)^{3/2} + C$

177. $\int x^2 \sqrt{a + bx}\, dx = \dfrac{2\,(8a^2 - 12abx + 15b^2x^2)}{105\ b^3} (a + bx)^{3/2} + C$

178. $\int \sqrt{\dfrac{x + a}{x + b}}\, dx = \sqrt{(x + a)(x + b)} + (a - b) \ln (\sqrt{x + a}$

$\qquad + \sqrt{x + b}) + C$

179. $\int \sqrt{\dfrac{a - x}{b + x}}\, dx = \sqrt{(a - x)(b + x)} + (a + b) \sin^{-1} \sqrt{\dfrac{b + x}{a + b}} + C$

Expressions involving algebraic equations.

180. $\int \dfrac{dx}{a^2 + x^2} = \dfrac{1}{a} \tan^{-1} \dfrac{x}{a} + C$

181. $\int \dfrac{dx}{x^2 - a^2} = \dfrac{1}{2a} [\ln (x - a) - \ln (x + a)] = \dfrac{1}{2a} \ln \dfrac{x - a}{x + a} + C$

182. $\int \sqrt{2\, ax - x^2}\, dx = \dfrac{1}{2} \left[(x - a) \sqrt{2\, ax - x^2} + a^2 \sin^{-1} \dfrac{x - a}{a}\right] + C$

183. $\int \dfrac{dx}{\sqrt{2ax - x^2}} = \sin^{-1} \dfrac{x - a}{a} + C$

184. $\int \dfrac{dx}{x\,(ax + b)} = \dfrac{1}{b} \ln \dfrac{x}{ax + b} + C$

Expressions involving trigonometric and exponential expressions.

185. $\int \sin^2 ax\, dx = \dfrac{1}{a} \left[\dfrac{ax}{2} - \dfrac{1}{4} \sin 2ax\right] + C$

186. $\int \sin^3 ax\, dx = \dfrac{1}{a} \left[-\cos ax + \dfrac{1}{3} \cos^3 ax\right] + C$

187. $\int \sin^4 ax\, dx = \dfrac{1}{a} \left[\dfrac{3\, ax}{8} - \dfrac{1}{4} \sin 2ax + \dfrac{1}{32} \sin 4ax\right] + C$

188. $\int \cos^2 ax\, dx = \dfrac{1}{a} \left[\dfrac{ax}{2} + \dfrac{\sin 2ax}{4}\right] + C$

189. $\int \cos^3 ax\, dx = \dfrac{1}{a} \left[\sin ax - \dfrac{1}{3} \sin^3 ax\right] + C$

190. $\int \cos^4 ax\, dx = \dfrac{1}{a} \left[\dfrac{3ax}{8} + \dfrac{1}{4} \sin 2ax + \dfrac{1}{32} \sin 4ax\right] + C$

191. $\int \sin^2 ax \cos^2 ax\, dx = \dfrac{1}{a} \left[\dfrac{ax}{8} - \dfrac{1}{32} \sin 4ax\right] + C$

192. $\int e^{ax} \sin px\, dx = \dfrac{e^{ax}(a \sin px - p \cos px)}{a^2 + p^2} + C$

193. $\int e^{ax} \cos px\, dx = \dfrac{e^{ax}(a \cos px + p \sin px)}{a^2 + p^2} + C$

194. $\int \sin^2 x\, dx = -\dfrac{1}{4} \sin 2x + \dfrac{1}{2} x + C = -\dfrac{1}{2} \sin x \cos x + \dfrac{1}{2} x + C$

195. $\int \cos^2 x\, dx = \dfrac{1}{4} \sin 2x + \dfrac{1}{2} x + C = \dfrac{1}{2} \sin x \cos x + \dfrac{1}{2} x + C$

Miscellaneous expressions.

196. $\int x \sin ax\, dx = \dfrac{1}{a^2} \sin ax - \dfrac{1}{a} x \cos ax + C$

197. $\int x \cos ax \, dx = \frac{1}{a^2} \cos ax + \frac{1}{a} x \sin ax + C$

198. $\int \sin^{-1} ax \, dx = x \sin^{-1}ax + \frac{1}{a} \sqrt{1 - a^2x^2} + C$

199. $\int x \sin^{-1}ax \, dx = \frac{2a^2x^2 - 1}{4a^2} \sin^{-1}ax + \frac{x}{4a} \sqrt{1 - a^2x^2} + C$

200. $\int \ln x \, dx = x \ln x - x + C$

201. $\int x \ln x \, dx = \frac{x^2}{2} \ln x - \frac{x^2}{4} + C$

202. $\int xe^{ax} \, dx = \frac{e^{ax}}{a^2} (ax - 1) + C$

203. $\int x^2 e^{ax} \, dx = \frac{e^{ax}}{a^3} (a^2x^2 - 2ax + 2) + C$

204. $\int e^{ax} \cos bx \, dx = \frac{a \cos bx + b \sin bx}{a^2 + b^2} e^{ax} + C$

205. $\int e^{ax} \sin bx \, dx = \frac{a \sin bx - b \cos bx}{a^2 + b^2} e^{ax} + C$

206. $\int x^n e^{ax} dx = \frac{x^n e^{ax}}{a} \left[1 - \frac{n}{ax} + \frac{n(n-1)}{a^2x^2} - \dots \pm \frac{n!}{a^n x^n} \right] + C$

Axioms of differential calculus.

1. The first differential coefficient of the sine of an angle is its cosine.
2. The first differential coefficient of the cosine of an angle is its sine with an opposite sign.
3. The first differential coefficient of the sum of any number of simple functions is equal to the sum of the first differential coefficients of each of the functions considered separately.
4. The first differential coefficient of a function composed of different particular functions, will be the sum of the first differential coefficients of each of these functions considered separately and independent of each other.
5. When the variable of a function consists of two parts, the differential coefficient will be the same to whichever part of the variation be ascribed.
6. The first derived function of f(x) is represented by f'(x), the second differential coefficient by f''(x), the third by f'''(x), etc.
7. When a fraction $\frac{P}{Q}$ both of whose terms are functions of x becomes $\frac{0}{0}$ when a particular value is assigned to the variable as $x = a$, then $(x - a)$ is a common factor both of numerator and denominator.
8. When the variable upon which any proposed function depends passes through all degrees of magnitude, the different values of the function form first an increasing and then a decreasing series, or vice versa.
9. When y is a maximum or minimum, $(y' - y)$ and $(y_1 - y)$ both have the same sign.

10. When an equation is reduced between x and y to the form $y = f(x)$, x is called the independent variable and y the dependent variable.
11. Curves which admit of asymptotes are divided into hyperbolic and parabolic. Hyperbolic are those which admit of a rectilinear asymptote; parabolic those which do not.
12. All hyperbolic curves are involved in the class $y = Ax + Bax^{-\alpha} + bx^{-\beta} + \dots$
13. A tangent to a curve, when the point of contact is removed to an infinite distance, becomes an asymptote.
14. When the limits of X and Y are finite, they determine a rectilinear asymptote.
15. When X has a limit and Y is infinite, then there is an asymptote parallel to the axis Ay at the distance X from the origin.
16. When Y has a limit and X none, there is an asymptote parallel to the axis Ax at the distance Y from the origin.
17. When both X and Y are infinite, the curve does not admit of a rectilinear asymptotote and the same if the values be impossible.
18. If $X = 0$ and $Y = 0$, the asymptote passes through the origin and its direction is found by determining the values $\frac{dy}{dx}$ when x becomes infinite.
19. When two curves have the first two terms of their developments equal, they osculate and have contact of the first order.
20. When the second differential coefficients are equal in each in the principles of contact, they have a contact of the second order.
21. When n differential coefficients are equal in each, they have a contact of the nth order.
22. Every osculating circle is of the second order and the first and second differential coefficients obtained by differentiating the equations to the curve and the circle will be identical.
23. When the radius of curvature at any point of the cycloid is taken, it is equal to twice the normal.
24. Every contact of the circle of curvature at the points of greatest and least curvature is mutually of the third order.
25. Every curve is equally convexed or concaved towards the axis of the abscissas, according as $\frac{d^2y}{dx^2}$ is positive or negative.
26. All equations of the form $A + x = B + y$ where A and B are constant quantities and x and y are susceptible of all degrees of magnitude, then A equals B and x equals y.
27. When a variable quantity by being continually increased or continually diminished, approaches towards a fixed quantity and approaches nearer to this quantity than any assignable difference, but never reaches or becomes equal to it, then that fixed quantity is called the limit of the variable quantity.

28. All circles are equal to other circles if the limits of the areas of the inscribed and circumscribed polygons are equal.
29. When y is mutually a function of x, the first differential coefficient of y is considered as the limit of the ratio of the increment of the function to the increment of the variable.
30. When a quantity x is infinitely small relative to a finite magnitude a, the square of x or x^2 is infinitely small relative to x.
31. Each quantity dy is equal to the differential of y.
32. Each quantity dx is equal to the differential of x.
33. The differential of y is always equal to the first differential coefficient of y, multiplied by the differential of x.

Axioms of integral calculus.
1. Every integral of the sum of any number of functions is equal to the sum of the integrals of the individual terms.
2. Every integral of a function raised to any power is equal to the differential of the function.
3. Every integral whose numerator is the differential of its denominator is equal to the logarithim of the denominator.
4. The integral of every fraction whose denominator is a radical of the second degree and whose numerator is the differential of the quantity under the radical sign, is equal to twice that radical.
5. Differentials resolved into two factors can be equally integrated by parts and are mutual to the whole integral required.
6. The integral sign $\int$ is equal to a stylized S standing for summation.
7. All integrals are found by an inverse differentiation process and are equal to the functions of its derivatives.
8. Every whole rational fraction of the form $\frac{P}{Q}$ is equally decomposed into others:

$$\frac{A}{x - a}, \frac{A}{(x - a)^n}, \frac{Ax + B}{x^2 + px + q}, \frac{Ax + B}{(x^2 + px + q)^n}, \text{etc.}$$

9. All areas of parabolas are equal to $\frac{2}{3}$xy and are equal to $\frac{2}{3}$ of a circumscribing rectangle.
10. All areas of quadrants are equal to $\frac{1}{4}\pi ab$.
11. When the derivative of distance equals the velocity, the distance equals the integral of the velocity.
12. All inverses of partial derivatives are equal to the partial integrals.
13. All improper integrals which have equal finite limits are equal to their convergents.
14. All improper integrals which have no finite limits have integrals equal to their divergents.

Example of general equations.

Given $y = 2x^2 + \sqrt{x} - \frac{4}{x} + 3$, determine $\frac{dy}{dx}$.

Simplifying:

$$y = 2x^2 + x^{1/2} - \frac{4}{x} + 3$$

Using Eq. 3:

$$\frac{dy}{dx} = \frac{d}{dx}(2x^2) + \frac{d}{dx}(x^{1/2}) - \frac{d}{dx}(4x^{-1}) + \frac{d}{dx}(3)$$

Using Eqs. 1, 4 and 8:

$$\frac{dy}{dx} = 2\left(2x \frac{dx}{dx}\right) + \frac{1}{2}x^{-1/2} \frac{dx}{dx} - 4(-1)x^{-2} \frac{dx}{dx} + 0$$

Simplifying and using Eq. 2:

$$\frac{dy}{dx} = 4x \cdot 1 + \frac{1}{2\sqrt{x}} \cdot 1 + \frac{4}{x^2} \cdot 1$$

$$= 4x + \frac{1}{2\sqrt{x}} + \frac{4}{x^2}$$

Example of trigonometric functions.

Given $y = \sin 2x + \cos 3x$, determine $\frac{dy}{dx}$.

$$\frac{dy}{dx} = \frac{d}{dx} \sin 2x + \frac{d}{dx} \cos 3x$$

$$= \cos 2x \frac{d}{dx}(2x) - \sin 3x \frac{d}{dx}(3x)$$

Using Eqs. 22 and 23:

$$\frac{dy}{dx} = (\cos 2x)(2) - (\sin 3x)(3)$$

$$= 2 \cos 2x - 3 \sin 3x$$

Example of inverse trigonometric functions.

Given $y = \tan^{-1}(2x - 1)$, determine $\frac{dy}{dx}$.

Using Eq. 30:

$$\frac{dy}{dx} = \frac{d(2x - 1)}{1 + (2x - 1)^2}$$

$$= \frac{2dx}{1 + 4x^2 - 4x + 1}$$

$$= \frac{2dx}{2 + 4x^2 - 4x}$$

$$= \frac{dx}{2x^2 - 2x + 1}$$

Example of exponential functions.

Given $y = 2^{3x}$, determine $\frac{dy}{dx}$.

Let $u = 3x$, then $y = 2^u$
Using Eq. 34:

$$\frac{dy}{dx} = 2^u \ln 2 \frac{du}{dx}$$

$$= 2^{3x} \ln 2 \frac{d}{dx}(3x)$$
$$= 3(2^{3x} \ln 2)$$

Example of miscellaneous expressions.

Given $y = 3 \csc^3 \frac{x}{3}$, determine $\frac{dy}{dx}$

$$\frac{dy}{dx} = 3 \frac{d}{dx} \csc^3 \frac{x}{3}$$

Differentiating as a power:

$$\frac{dy}{dx} = 3\left(3 \csc^2 \frac{x}{3}\right) \frac{d}{dx} \csc \frac{x}{3}$$

Using Eq. 44 and letting $h = 1$:

$$\frac{dy}{dx} = 9 \csc^2 \frac{x}{3}\left(-\csc \frac{x}{3} \cot \frac{x}{3}\right)\frac{1}{3}$$
$$= -3 \csc^3 \frac{x}{3} \cot \frac{x}{3}$$

Example of first differential coefficient of functions of one variable.

Determine the first differential coefficient of $\sin x$.
Using Eq. 57:
Let $y = \sin x$
Let x become $x + h$ and y become y'
$\quad y' = \sin(x + h)$
$\quad y' = \sin x \cos h + \sin h \cos x$
Substituting for $\sin h$ and $\cos h$:

$$\sin x \left(1 - \frac{h^2}{1 \cdot 2} + \frac{h^4}{1 \cdot 2 \cdot 3 \cdot 4} - \cdots\right) +$$
$$\cos x \left(\frac{h}{1} - \frac{h^3}{1 \cdot 2 \cdot 3} + \frac{h^5}{1 \cdot 2 \cdot 3 \cdot 4 \cdot 5} - \cdots\right)$$

Arranging according to powers of h:

$$\sin x + \cos x \cdot h - \frac{\sin x}{1 \cdot 2}h^2 - \frac{\cos x}{1 \cdot 2 \cdot 3}h^3 + \cdots$$
$$\frac{dy}{dx} = \cos x$$

Example of inverse functions.

Given $y = \operatorname{cosec}^{-1} \frac{\sqrt{1 + x^2}}{x}$, determine $\frac{dy}{dx}$.

Let $y = \operatorname{cosec}^{-1} \frac{\sqrt{1 + x^2}}{x}$

Let $Z = \frac{\sqrt{1 + x^2}}{x}$

Using Eq. 82:
$\quad y = \operatorname{cosec}^{-1} Z$ or $Z = \operatorname{cosec} y$

$$\frac{dy}{dx} = \left(\frac{dy}{dx}\right)\left(\frac{dZ}{dx}\right)$$
$$= -\frac{1}{Z\sqrt{Z^2 - 1}} \cdot -\frac{1}{x^2\sqrt{1 + x^2}}$$
$$= \frac{1}{1 + x^2}$$

Example of elementary forms of the integral equations.

Determine the value of $\int (x^2 - 2\sqrt{x} + 3)\,dx$

$$\int (x^2 - 2\sqrt{x} + 3)\,dx = \int (x^2\,dx - 2x^{1/2}\,dx + 3dx)$$

Using Eq. 102:

$$\int (x^2 - 2\sqrt{x} + 3)\,dx = \int x^2\,dx - \int 2x^{1/2}\,dx + \int 3dx$$

Using Eq. 103:

$$\int (x^2 - 2\sqrt{x} + 3)\,dx = \int x^2\,dx - 2\int x^{1/2}\,dx + 3\int dx$$
$$= \frac{x^3}{3} - 2\frac{x^{3/2}}{\frac{3}{2}} + 3x + C$$
$$= \frac{x^3}{3} - \frac{4}{3}x^{3/2} + 3x + C$$

Example of fundamental integrals.

Determine the value of $\int \left(\frac{1}{3x^2} - \frac{2}{x}\right)dx$

Using Eqs. 111 and 112:

$$\int \left(\frac{1}{3x^2} - \frac{2}{x}\right)dx = \int \left(\frac{x^{-2}}{3}dx - 2\frac{dx}{x}\right)$$
$$= \int \frac{x^{-2}}{3}dx - \int 2\frac{dx}{x}$$
$$= \frac{1}{3}\int x^{-2}dx - 2\int \frac{dx}{x}$$
$$= \frac{1}{3}\left(\frac{x^{-1}}{-1}\right) - 2\ln x + C$$

Simplifying:

$$= -\frac{1}{3x} - 2\ln x + C$$

Example of expressions involving $\sqrt{a^2 - x^2}$ and $\dfrac{du}{\sqrt{a^2 + u^2}}$.

Determine the value of $\int \dfrac{x^3 dx}{\sqrt{a^2 - x^2}}$

Using Eq. 141:

Let $du = \dfrac{x^3 dx}{\sqrt{a^2 - x^2}}$

$$u = \int x^2 \frac{x\,dx}{\sqrt{a^2 - x^2}}$$
$$= -x^2\sqrt{a^2 - x^2} + 2\int x\,dx \sqrt{a^2 - x^2}$$
$$= -x^2\sqrt{a^2 - x^2} + 2a^2\int \frac{x\,dx}{\sqrt{a^2 - x^2}} - 2u$$
$$= -\frac{x^2}{3}\sqrt{a^2 - x^2} + \frac{2a^2}{3}\int \frac{x\,dx}{\sqrt{a^2 - x^2}}$$
$$= -\frac{x^2}{3}\sqrt{a^2 - x^2} - \frac{2a^2}{3}\sqrt{a^2 - x^2} + C$$
$$= -\sqrt{a^2 - x^2}\left\{\frac{x^2}{3} + \frac{2a^2}{3}\right\} + C$$

Example of expressions involving $\sqrt{x^2 \pm a^2}$ **and** $x\sqrt{x^2 - a^2}$.

Determine the value of $\int \dfrac{x^3\,dx}{\sqrt{x^2 \pm a^2}}$.

Using Eq. 154:

$$\int \frac{x^2\,dx}{\sqrt{x^2 \pm a^2}} = \frac{1}{3}(x^2 \mp 2a^2)\sqrt{x^2 \pm a^2} + C$$

Example of expressions involving $\sqrt{a + bx}$ **and** $\sqrt{\dfrac{a-x}{b+x}}$.

Determine the value of $\int \dfrac{x^3\,dx}{(a + bx^2)^{3/2}}$.

Using Eqs. 170 and 171:

$$\text{Let } du = \frac{x^3\,dx}{(a + bx^2)^{3/2}}$$

$$u = \int x^2 \cdot x\,dx\,(a + bx^2)^{-3/2}$$

$$\int \frac{x^3\,dx}{(a + bx^2)^{3/2}} = -\frac{x^2}{b} \cdot \frac{1}{\sqrt{a + bx^2}} + \frac{a}{b}\int \frac{x\,dx}{\sqrt{a + bx^2}}$$

$$= -\frac{x^2}{b} \cdot \frac{1}{\sqrt{a + bx^2}} + \frac{a}{b^2}\sqrt{a + bx^2}$$

$$= \frac{1}{\sqrt{a + bx^2}}\left\{\frac{x^2}{b} + \frac{2a}{b^2}\right\}$$

Example of expressions involving algebraic equations.

Determine the value of $\int \dfrac{dx}{4x^2 + 9x}$.

$$\int \frac{dx}{4x^2 + 9x} = \frac{1}{4}\int \frac{dx}{x^2 + \dfrac{9}{4}x}$$

$$= \frac{1}{4}\int \frac{dx}{\left(x^2 + \dfrac{9}{4}x + \dfrac{81}{64}\right) - \dfrac{81}{64}}$$

$$= \frac{1}{4}\int \frac{dx}{\left(x + \dfrac{9}{8}\right)^2 - \left(\dfrac{9}{8}\right)^2}$$

Let $u = x + \dfrac{9}{8}$ and $du = dx$

Using Eq. 181:

$$\int \frac{dx}{4x^2 + 9x} = \frac{1}{4}\int \frac{du}{u^2 - \left(\dfrac{9}{8}\right)^2}$$

$$= \frac{1}{9}\ln\frac{u - \dfrac{9}{8}}{u + \dfrac{9}{8}} + C$$

$$= \frac{1}{9}\ln\frac{8u - 9}{8u + 9} + C$$

$$= \frac{1}{9}\ln\frac{8\left(x + \dfrac{9}{8}\right) - 9}{8\left(x + \dfrac{9}{8}\right) + 9} + C$$

$$= \frac{1}{9}\ln\frac{8x}{8x + 18} + C$$

$$= \frac{1}{9}\ln\frac{4x}{4x + 9} + C$$

Example of expressions involving trigonometric and exponential expressions.

Determine the value of $\int \sin^4 x\,dx$.

Using Eq. 187:

$$\int \sin^4 x\,dx = \int (\sin^2 x)^2\,dx$$

$$= \int \left(\frac{1 - \cos 2x}{2}\right)^2 dx$$

$$= \frac{1}{4}\int (1 - 2\cos 2x + \cos^2 2x)dx$$

$$= \frac{1}{4}\int \left(1 - 2\cos 2x + \frac{1 + \cos 4x}{2}\right)dx$$

$$= \frac{1}{4}\int \left(1 - 2\cos 2x + \frac{1}{2} + \frac{1}{2}\cos 4x\right)dx$$

$$= \frac{1}{4}\int \left(\frac{3}{2} - 2\cos 2x + \frac{1}{2}\cos 4x\right)dx$$

$$= \frac{1}{4}\left(\frac{3}{2}x - \sin 2x + \frac{1}{8}\sin 4x\right) + C$$

Example of miscellaneous expressions.

Determine the value of $\int xe^{2x}\,dx$

Using Eq. 202:

Let $\mu = x$ and $dv = e^{2x}\,dx$

Then $du = dx$ and $v = \dfrac{1}{2}e^{2x}$

By substituting in the following equation:

$$\int \mu\,dv = dv$$
$$= \mu v - \int v\,du$$

Solving for our answer:

$$\int xe^{2x}\,dx = \frac{xe^{2x}}{2} - \int \left(\frac{1}{2}e^{2x}\right)dx$$

$$= \frac{xe^{2x}}{2} - \frac{1}{4}e^{2x} + C$$

Chart for cubic equations

Introduction.

Frequently in engineering calculations it is necessary to solve a cubic equation. The following charts may be applied to quickly determine how many real roots of a given cubic equation exist. Also, if the cubic equation has more than one real root, the roots themselves may be approximated quickly and easily. Even in the "irreducible case" all three roots can be located on the charts without complicated computation or necessity of operating with complex numbers. The following charts do not assist in locating roots in the case where there are one real and two conjugate complex roots.

Use of charts.

Given a cubic equation in the form:

$$y^3 + Py^2 + qy + r = 0$$

Where P, q and r are real numbers.

Step 1.

Reduce the above equation to:

$$x^3 + ax + b = 0 \qquad (1)$$

This may be done by any of the following conventional methods:

(1) Transformation letting $x = y - \left(\dfrac{P}{3}\right)$

(2) Successive synthetic divisions by $\left(-\dfrac{P}{3}\right)$

(3) Direct application of the general equations:

$$a = q - 3\left(\frac{P}{3}\right)^2$$

$$b = 2\left(\frac{P}{3}\right)^3 - q\left(\frac{P}{3}\right) + r$$

Note: If a is positive then it directly follows that the cubic equation has one real root and two conjugate complex roots.

Step 2.

Refer to the Discriminate Chart and locate a on the y scale and b on the x scale without regard to their sign.

Step 3.

Observe the relationship of the points a and b to the curve.

Case 1: If a and b lie above the curve, the cubic equation has one real and two conjugate complex roots.

Case 2: If a and b fall on the curve, the cubic equation has a multiple root. In this case the value of the roots may be found by projecting the a value horizontally to intersect the curves on the Root Chart. Read the values of x at each intersection of the curves. The first curve intersected (multiple root) gives the root of multiplicity 2 and the second curve intersected (simple root) gives the simple root. Sign of the root is determined from the Sign Table.

Case 3: If a and b fall below the curve, the cubic equation has three real and distinct roots. In this case the value of the roots may be found by reading the value of b that corresponds to a on the Discriminate Chart and let this value be b_0. Divide b from the cubic equation by b_0 and the resulting quotient will always be between 0 and 1 and referred to as $\cos\phi$. Referring to the Cosine Chart, locate $\cos\phi$ on the y axis, project its value horizontally to intersect the curves on the Cosine Chart, read the three values of x that correspond to the value of $\cos\phi$ and let these be called alpha, beta and gamma. Refer to the Root Chart and read the values of x which would be the simple root in case 2. Let this be x_0.

The three roots are:

$$x_1 = x_0 \text{(alpha)}$$
$$x_2 = x_0 \text{(beta)}$$
$$x_3 = x_0 \text{(gamma)}$$

Note:

If in arriving at the reduced cubic equation, the values of a or b fall outside the ranges plotted, use a simple transformation such as $z = 0.5x$. If a or b are small, use a transformation such as $z = 3x$.

Example 1:

Solve the following cubic equation that is in reduced form:

$$x^3 - 30x + 64 = 0$$

Solution 1:

From inspection of this equation (compared to Eq. 1), the following values are obtained:

$$a = -30$$
$$b = +64$$

On the Discriminate Chart locate 30 on the a scale and 64 on the b scale and note where they intersect the discriminate curve. Since a and b fall on the curve, we have a Case 2 cubic equation. From the intersection of a and b on the discriminate curve, construct a horizontal line to intersect the root curves on the Root Chart. There is a double root at 3.2 and a simple root at 6.3. The final signs determined from the Sign Table are:

$$x = +3.2$$
$$x = -6.3$$

Example 2:

Solve the following cubic equation that is in reduced form:

$$x^3 - 96x - 256 = 0$$

Solution 2:

From inspection of this equation (compared to Eq. 1), the following values are obtained:

$$a = -96$$
$$b = -256$$

On the Discriminate Chart locate 96 on the a scale and 256 on the b scale and note that they intersect below the discriminate curve. Since a and b intersect below the discriminate curve, we have a Case 3 cubic equation. On the discriminate curve, the value of b that corresponds to a is 365, thus $b_0 = 365$. Next, determine $\cos\phi$:

$$\frac{b}{b_0} = \frac{256}{365} = 0.70 = \cos\phi$$

Enter the Cosine Chart at 0.70 and construct a horizontal line to intersect the alpha (α), beta (β) and gamma (γ) curves. Read the following values (signs determined from the Sign Table):

$$\alpha = -0.25$$
$$\beta = -0.71$$
$$\gamma = +0.97$$

Referring to the Root Chart read the value of x which would be the simple root, $x_0 = 11.3$. Thus the roots are:

$$x_1 = x_0 \text{ (alpha)}$$
$$x_1 = 11.3\,(-0.25)$$
$$x_1 = -2.8$$

$$x_2 = x_0 \text{ (beta)}$$
$$x_2 = 11.3\,(-0.71)$$
$$x_2 = -8.0$$

$$x_3 = x_0 \text{ (gamma)}$$
$$x_3 = 11.3\,(+0.97)$$
$$x_3 = +10.9$$

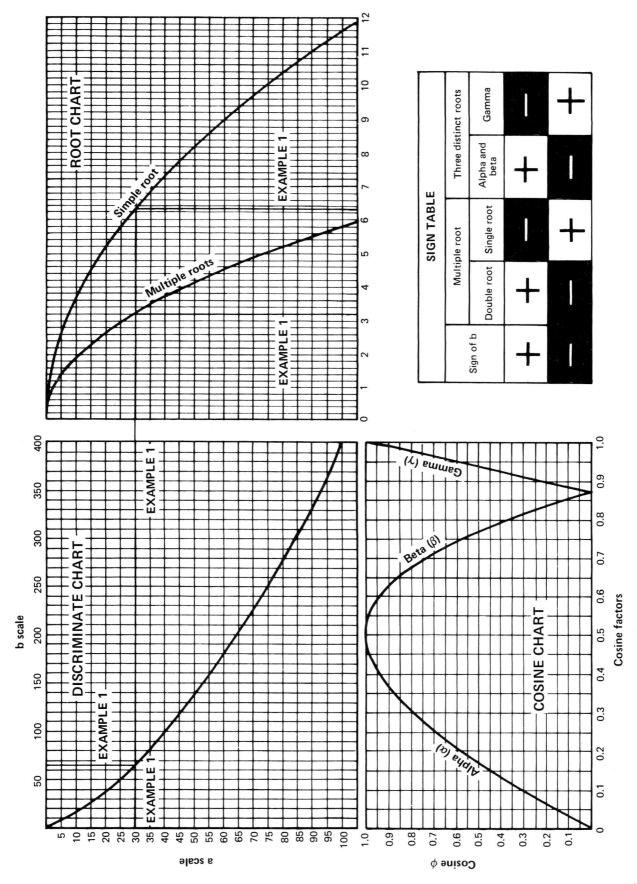

Equations containing π

Introduction.

In the solution of engineering problems it is often necessary to solve equations containing the constant π. This article contains the most commonly used equations that contain π. In each case the number given at the right of the equation is the numerical value of the function of π enclosed by the parenthesis.

Nomenclature:

C = Circumference, inches
D = Diameter, inches
R = Radius, inches
A = Cross-sectional area, sq in
A_c = Area under a cycloid, sq in
S = Surface area, sq in
V = Volume, cu in
I_p = Polar moment of inertia, inches^4
I_T = Transverse moment of inertia, inches^4
Z_p = Polar section modulus, inches^3
Z_T = Transverse section modulus, inches^3
$\overline{y}$ = Centroid, inches

Example 1:

Given a circle that has a diameter of 3.0 inches, determine its area.

Solution 1:

Selecting the proper area equation we have:

$$A = \left(\frac{\pi}{4}\right) D^2$$

From this article we know

$$\left(\frac{\pi}{4}\right) = 0.78540$$

Thus we solve the equation:

$$A = (0.78540)(3.0^2)$$

$$A = 7.06 \text{ sq in}$$

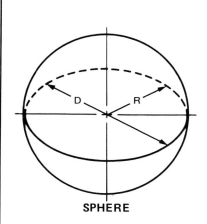

SPHERE

$$V = \left(\frac{4\pi}{3}\right)R^3 \qquad 4.18879$$

$$V = \left(\frac{\pi}{6}\right)D^3 \qquad 0.52360$$

$$R = \left(\sqrt[3]{\frac{3}{4\pi}}\right)\sqrt[3]{V} \qquad 0.62035$$

$$D = \left(\sqrt[3]{\frac{6}{\pi}}\right)\sqrt[3]{V} \qquad 1.24070$$

$$S = (4\pi)r^2 \qquad 12.56637$$

$$S = (\pi)D^2 \qquad 3.14159$$

$$R = \left(\frac{1}{2\sqrt{\pi}}\right)\sqrt{S} \qquad 0.28209$$

$$D = \left(\frac{1}{\pi}\right)\sqrt{S} \qquad 0.31831$$

INTEGRALS

$$\int_0^1 \frac{\text{Log } x}{x-1}\,dx = \left(\frac{\pi^2}{6}\right) = 1/1^2 + 1/2^2 + 1/3^2$$
$$1.64493$$

$$\int_0^1 \frac{\text{Log } x}{x^2-1}\,dx = \left(\frac{\pi^2}{8}\right) = 1/1^2 + 1/3^2 + 1/5^2$$
$$1.23370$$

$$\int_0^1 \text{Log}\left(\frac{1+x}{1-x}\right)\frac{dx}{x} = \left(\frac{\pi^2}{4}\right) \qquad 2.46740$$

$$\int_0^1 \frac{\text{Log } x}{1+x}\,dx = -\left(\frac{\pi^2}{12}\right) \qquad 0.82247$$

$$\int_0^1 \left(\text{Log }\frac{1}{x}\right)^{1/2}dx = \left(\frac{\sqrt{\pi}}{2}\right) \qquad 0.88623$$

$$\int_0^1 \left(\text{Log }\frac{1}{x}\right)^{-1/2}dx = (\sqrt{\pi}) \qquad 1.77245$$

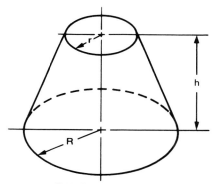

FRUSTUM OF A RIGHT-CIRCULAR CONE

$$V = \left(\frac{\pi}{3}\right)h(R^2 + Rr + r^2) \qquad 1.04720$$

$$V = \left(\frac{\pi}{12}\right)h(D^2 + Dd + d^2) \qquad 0.26180$$

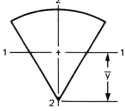

ONE-SIXTH SECTOR

$$\overline{y} = \left(\frac{2}{\pi}\right)R \qquad 0.63662$$

$$I_{1-1} = \left(\frac{\pi}{24} + \frac{\sqrt{3}}{16} - \frac{2}{3\pi}\right)R^4 \qquad 0.02695$$

$$I_{2-2} = \left(\frac{\pi}{24} - \frac{\sqrt{3}}{16}\right)R^4 \qquad 0.02265$$

AREA UNDER A CYCLOID

$$A_c = (3\pi)R^2 \qquad 9.42478$$

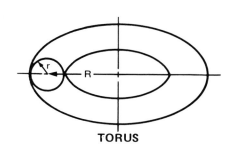

TORUS

$$V = (2\pi^2)Rr^2 \qquad 19.73921$$

$$V = \left(\frac{\pi^2}{4}\right)Dd^2 \qquad 2.46740$$

$$S = (4\pi^2)Rr \qquad 39.47842$$

$$S = (\pi^2)Dd \qquad 9.86960$$

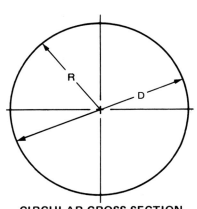

CIRCULAR CROSS-SECTION

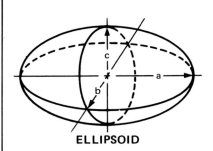

$$\bar{y} = \left(\frac{4}{3\pi}\right)R \qquad 0.42441$$

$$I_{1-1} = \left(\frac{9\pi^2 - 64}{72\pi}\right)R^4 \qquad 0.10976$$

$$I_{2-2} = \left(\frac{\pi}{8}\right)R^4 \qquad 0.39270$$

SEMICIRCLE

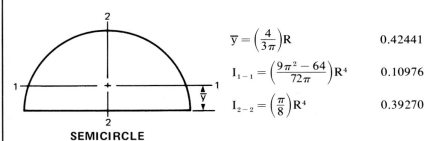

$$V = \left(\frac{4\pi}{3}\right)abc \qquad 4.18879$$

ELLIPSOID

$C = (\pi)D$	3.14159
$C = (2\pi)R$	6.28319
$D = (1/\pi)C$	0.31831
$R = (1/2\pi)C$	0.15915
$A = (\pi)R^2$	3.14159
$A = (\pi/4)D^2$	0.78540
$D = (\sqrt{4/\pi})\sqrt{A}$	1.12838
$R = (\sqrt{1/\pi})\sqrt{A}$	0.56419
$I_T = (\pi/4)R^4$	0.78540
$I_T = (\pi/64)D^4$	0.04909
$I_P = (\pi/2)R^4$	1.57080
$I_P = (\pi/32)D^4$	0.09817
$D = \sqrt[4]{32/\pi}\ \sqrt[4]{I_P}$	1.78649
$D = \sqrt[4]{64/\pi}\ \sqrt[4]{I_T}$	2.12450
$R = \sqrt[4]{2/\pi}\ \sqrt[4]{I_P}$	0.89324
$R = \sqrt[4]{4/\pi}\ \sqrt[4]{I_T}$	1.06225
$Z_P = (\pi/2)R^3$	1.57080
$Z_P = (\pi/16)D^3$	0.19635
$D = \sqrt[3]{16/\pi}\ \sqrt[3]{Z_P}$	1.72051
$R = \sqrt[3]{2/\pi}\ \sqrt[3]{Z_P}$	0.86025
$Z_T = (\pi/4)R^3$	0.78540
$Z_T = (\pi/32)D^3$	0.09817
$D = \sqrt[3]{32/\pi}\ \sqrt[3]{Z_T}$	2.16770
$R = \sqrt[3]{4/\pi}\ \sqrt[3]{Z_T}$	1.08385

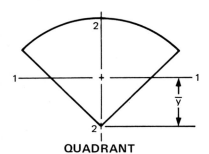

$$\bar{y} = \left(\frac{4\sqrt{2}}{3\pi}\right)R \qquad 0.60021$$

$$I_{1-1} = \left(\frac{\pi}{16} + \frac{1}{8} + \frac{8}{9\pi}\right)R^4 \qquad 0.60429$$

$$I_{2-2} = \left(\frac{\pi}{16} - \frac{1}{8}\right)R^4 \qquad 0.07135$$

QUADRANT

Definite integral nomogram

Introduction.

A definite integral from a geometrical point of view is the surface included between the x-axis, the integrated function and two vertical lines, plotted in the upper and lower integration limits.

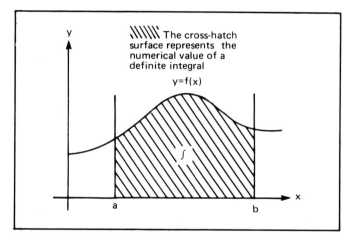

It can be mathematically expressed as follows:

$$\int_a^b f(x)dx = f(x)\Big|_a^b = f(b) - f(a) \qquad (1)$$

Eq. 1 states: the definite integral of a function f (x) equals the difference between the integral value in its upper limit (b) minus the integral value in its lower limit (a). The calculation of this surface under a given curve is a problem frequently encountered by engineers.

There are some elementary functions of special importance that are used in everyday calculations. The following 10 functions have been chosen and a Nomogram has been constructed that allows the quick calculation of the definite integral of these functions.

$Y = X$	$Y = \tan X$	$Y = \dfrac{1}{X}$
$Y = X^2$	$Y = \ln X$	
$Y = \sqrt{X}$	$Y = \sqrt{2X - X^2}$	$Y = \dfrac{1}{\sqrt{X}}$
$Y = \sin X$	$Y = e^x$	

Nomogram.

The Definite Integral Nomogram is constructed from three concentric circles. On these circles 12 scales are plotted. The two largest scales, on the upper and lower semicircles, are identical and represent the values a and b for lower and upper integration limits and are called scales 1 and 2, respectively.

The other 10 semicircular scales are plotted with the values of the appropriate integrals.

The use of the nomogram is very simple since it is known that the definite integral value equals the difference between the integral values in its upper and lower limits.

Procedure.

1. Connect with a ruler the known value of the lower integration limit on both scale 1 and scale 2. Read the value of f(a) on the intersection point of the ruler with the appropriate integral scale.

2. Connect with a ruler the known value of the upper integration limit on both scale 1 and scale 2. Read the value of f(b) on the intersection point of the ruler with the appropriate integral scale.

3. The final result is obtained by subtracting f(b) − f(a).

Comments.

1. The values of X in the trigonometric functions are expressed in radians. The following relations will be useful in using the Nomogram.

1 radian = 57 deg 17 min

$\frac{\pi}{2}$ radians = 90 deg

π radians = 180 deg

1 second = 0.0000048 radians

1 minute = 0.000291 radians	30 deg = 0.523599 radians
1 deg = 0.017453 radians	45 deg = 0.785398 radians
15 deg = 0.261799 radians	90 deg = 1.570796 radians

2. When the angle α equals 90 deg, the function tan α is infinite and the integration is impossible. Therefore, the integration on the nomogram is impossible when the angle reaches 90 deg since the function as well as the integral equals infinity. The integration may be performed only within the following limits:

0 to 89 deg (0 to 1.55 radians)
91 to 180 deg (1.59 to 3.14 radians)

It is therefore impossible to integrate a tangent function between 70 to 100 deg because of the above limitations.

3. When integrating the function $Y = \ln X$, it is possible to obtain a negative value of the integral. Its geometrical interpretation indicates that the surface lies under the X-axis.

Example 1:

What is the definite integral of the function $Y = X^2$ (What will be the surface under a parabola, limited by the vertical lines X = 1.0 and X = 3.0)?

$$\int_{1.0}^{3.0} X^2 dx = ?$$

Solution 1:

Connect with a ruler the a value on both scale 1 and scale 2 (a = 1.0). At the intersection of the ruler with the $\int X^2 dx$ scale, read the answer of a = 0.33. Connect with a ruler the b value on both scale 1 and scale 2 (b = 3.0). At the intersection of the ruler with the $\int X^2 dx$ scale, read the answer of b = 9.0. Subtract the following:

$$9.0 - 0.33 = 8.67$$
$$\int_{1.0}^{3.0} X^2 dx = 8.67$$

Example 2:

What is the definite integral of the function $Y = \ln X$ (What will be the surface under a logarithmic curve, limited by the vertical lines X = 1.0 and X = 3.0)?

$$\int_{1.0}^{3.0} \ln X dx = ?$$

Solution 2:

Connect with a ruler the a value on both scale 1 and scale 2 (a = 1.0). At the intersection of the ruler with the $\int \ln X dx$ scale, read the answer of a = −1. Connect with a ruler the b value on both scale 1 and scale 2 (b = 3.0). At the intersection of the ruler with the $\int \ln X dx$ scale, read the answer of b = 0.29. Subtract the following:

$$0.29 - (-1) = 0.29 + 1 = 1.29$$
$$\int_{1.0}^{3.0} \ln X dx = 1.29$$

FUNCTION	TYPE OF CURVE	INTEGRAL	SKETCH
$y = x$	STRAIGHT, OBLIQUE LINE	$\dfrac{x^2}{2}$	
$y = x^2$	PARABOLIC	$\dfrac{x^3}{3}$	
$y = \sqrt{x}$	INVERTED PARABOLIC	$\dfrac{2}{3}\sqrt{x^3}$	
$y = \sin x$	SINUSOIDAL	$-\cos x - 1$	
$y = \tan x$	TANGENTIAL	$-\ln \cos x$	

NOTE: THE CROSS-HATCH SURFACES REPRESENT THE NUMERICAL VALUES OF THE DEFINITE INTEGRALS

FUNCTION	TYPE OF CURVE	INTEGRAL	SKETCH
$y = \ln x$	LOGARITHMIC	$x \ln x - x$	
$y = \sqrt{2x - x^2}$	SEMICIRCULAR	$\dfrac{x-1}{2} \cdot \sqrt{2x - x^2} +$ arc sin $(1-x) - \dfrac{\pi}{2}$	
$y = e^x$	EXPONENTIAL	e^x	
$y = \dfrac{1}{x}$	HYPERBOLIC	$\ln x$	
$y = \dfrac{1}{\sqrt{x}}$	HYPERBOLA LIKE	$2\sqrt{x}$	

NOTE: THE CROSS-HATCH SURFACES REPRESENT THE NUMERICAL VALUES OF THE DEFINITE INTEGRALS

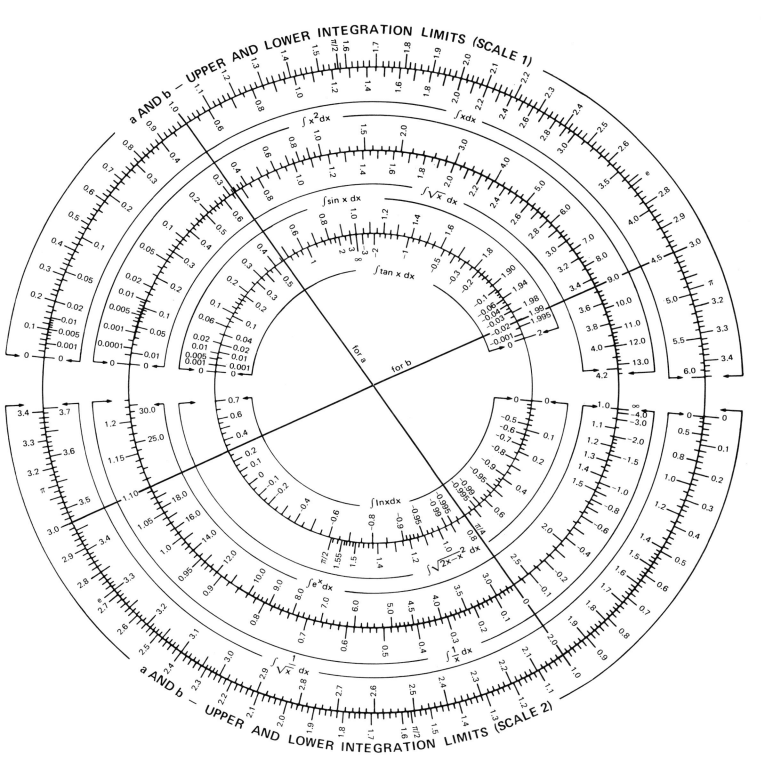

Nomogram for solving

$$c = \sqrt{a^2 + b^2}$$

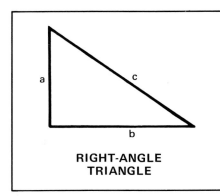

**RIGHT-ANGLE
TRIANGLE**

Introduction.
Frequently in engineering calculations the solution to the following equation is required:
$$c = \sqrt{a^2 + b^2}$$
Nomogram.
A nomogram can be used to expedite the solution of the above equation.
Nomenclature:
a = Side of triangle, inches
b = Side of triangle, inches
c = Hypotenuse, inches
Example 1:
Given a triangle that has an a = 9.0 inches and b = 12.0 inches, determine c.

Solution 1.
Construct a line from 9.0 on the a scale to 12.0 on the b scale and where this line intersects the c scale read the answer of 15.0 inches.

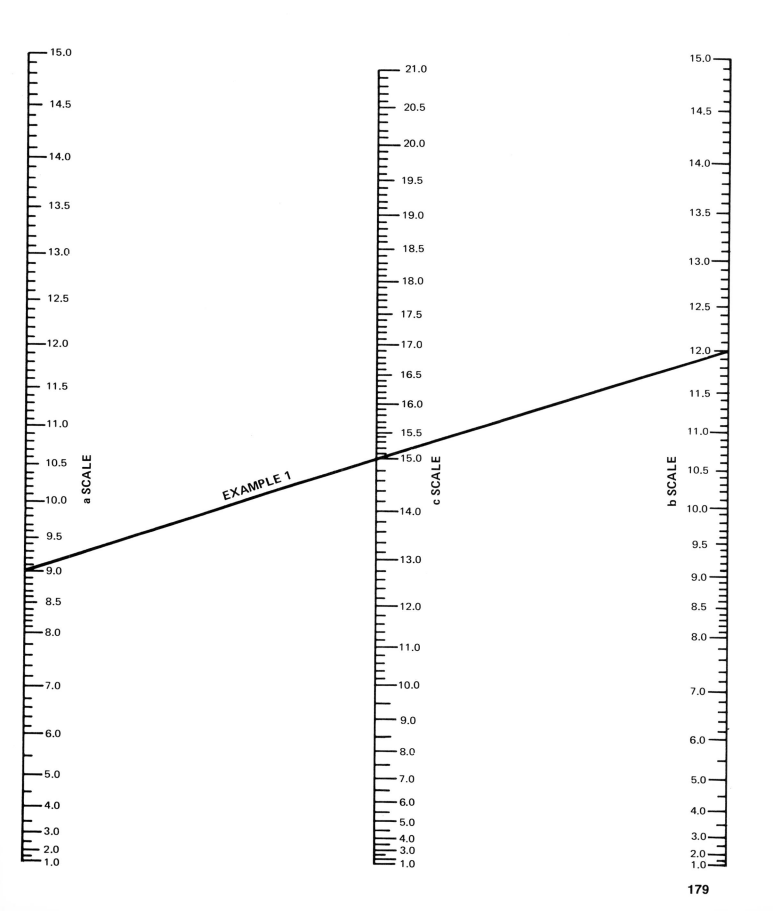

Nomogram for solving $Z = X^Y$

Introduction.

Frequently in engineering calculations the solution to the following equation is required:

$$Z = X^Y$$

Nomogram.

A nomogram can be used to expedite the solution of the above equation.

Example 1:

Solve $\sqrt[5]{0.00245}$

Solution 1:

Construct a line from 0.00245 on the X scale to 0.20 on the Y scale (1/5 = 0.20) and continue this line until it intersects the Z scale and read the answer of Z = 0.30.

Example 2:

Solve $14^{1.5}$

Solution 2:

Construct a line from 14 on the X scale to 1.5 on the Y scale and continue this line until it intersects the Z scale and read the answer of Z = 55.

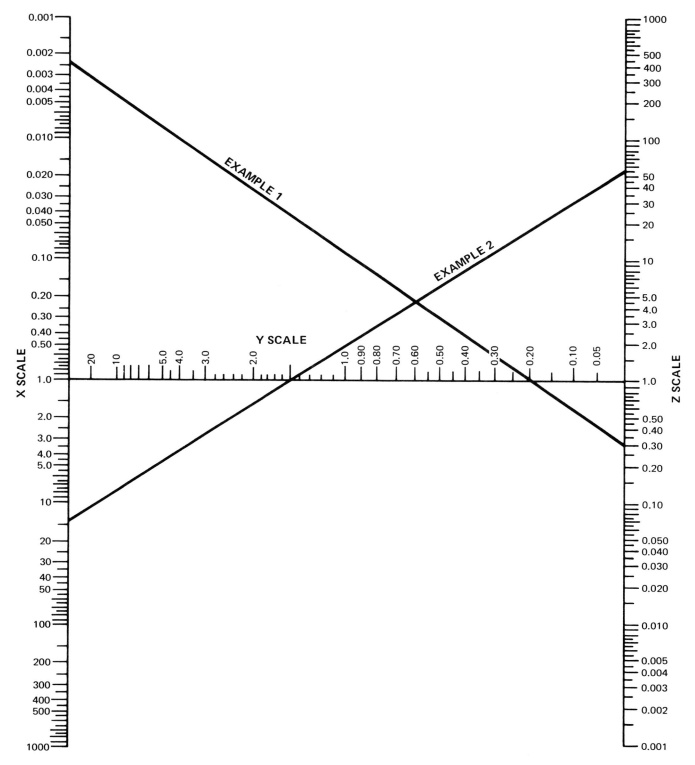

Circular cross-sections

Introduction.

Occasionally it is necessary to place a circular cross-section in the space resulting from a combination of plane and/or curved surfaces. The following tables give the equations for finding the diameter of the largest circle that will fit in each situation.

Example 1:

Given three equal cylinders (butted together), each with a diameter (d) of 1.5 inches, determine the largest diameter (D) that will fit between the circles.

Solution 1:

Going to the table for three equal cylinders (butted together), the following equation is obtained:

$$D = 0.1547d$$

Substituting into this equation:

$$D = 0.1547(1.5)$$
$$D = 0.232 \text{ inch}$$

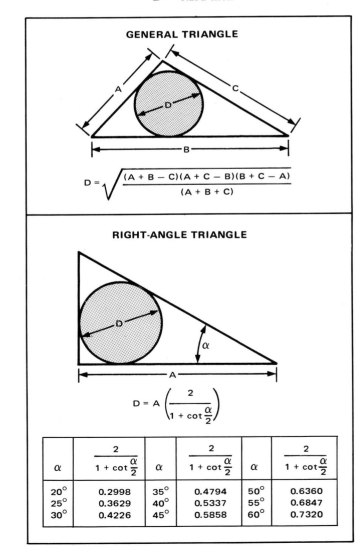

GENERAL TRIANGLE

$$D = \sqrt{\frac{(A + B - C)(A + C - B)(B + C - A)}{(A + B + C)}}$$

RIGHT-ANGLE TRIANGLE

$$D = A\left(\frac{2}{1 + \cot\frac{\alpha}{2}}\right)$$

α	$\dfrac{2}{1 + \cot\frac{\alpha}{2}}$	α	$\dfrac{2}{1 + \cot\frac{\alpha}{2}}$	α	$\dfrac{2}{1 + \cot\frac{\alpha}{2}}$
20°	0.2998	35°	0.4794	50°	0.6360
25°	0.3629	40°	0.5337	55°	0.6847
30°	0.4226	45°	0.5858	60°	0.7320

ISOSCELES TRIANGLE

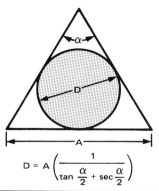

$$D = A\left(\dfrac{1}{\tan\frac{\alpha}{2} + \sec\frac{\alpha}{2}}\right)$$

α	$\dfrac{1}{\tan\frac{\alpha}{2} + \sec\frac{\alpha}{2}}$	α	$\dfrac{1}{\tan\frac{\alpha}{2} + \sec\frac{\alpha}{2}}$	α	$\dfrac{1}{\tan\frac{\alpha}{2} + \sec\frac{\alpha}{2}}$
20°	0.8391	55°	0.6067	90°	0.4142
25°	0.8025	60°	0.5773	95°	0.3888
30°	0.7672	65°	0.5486	100°	0.3639
35°	0.7332	70°	0.5205	105°	0.3394
40°	0.7001	75°	0.4931	110°	0.3153
45°	0.6681	80°	0.4663	115°	0.2914
50°	0.6370	85°	0.4400	120°	0.2679

TWO PLANE SURFACES AND A CYLINDER

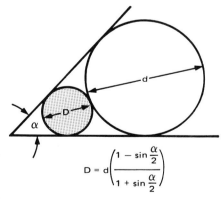

$$D = d\left(\dfrac{1 - \sin\frac{\alpha}{2}}{1 + \sin\frac{\alpha}{2}}\right)$$

α	$\dfrac{1 - \sin\frac{\alpha}{2}}{1 + \sin\frac{\alpha}{2}}$	α	$\dfrac{1 - \sin\frac{\alpha}{2}}{1 + \sin\frac{\alpha}{2}}$	α	$\dfrac{1 - \sin\frac{\alpha}{2}}{1 + \sin\frac{\alpha}{2}}$
20°	0.7041	55°	0.3682	90°	0.1716
25°	0.6442	60°	0.3333	95°	0.1512
30°	0.5888	65°	0.3010	100°	0.1325
35°	0.5376	70°	0.2710	105°	0.1152
40°	0.4903	75°	0.2432	110°	0.0994
45°	0.4465	80°	0.2174	115°	0.0850
50°	0.4059	85°	0.1936	120°	0.0718

TWO CYLINDERS AND A PLANE SURFACE

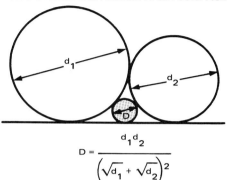

$$D = \dfrac{d_1 d_2}{\left(\sqrt{d_1} + \sqrt{d_2}\right)^2}$$

TWO EQUAL CYLINDERS AND A PLANE SURFACE

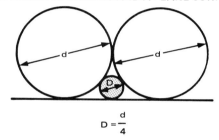

$$D = \dfrac{d}{4}$$

TWO EQUAL CYLINDERS AND A CYLINDRICAL SURFACE

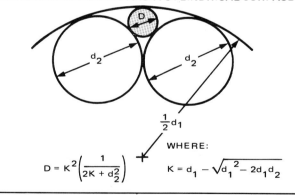

$$D = K^2\left(\dfrac{1}{2K + d_2}\right)$$

WHERE:

$$K = d_1 - \sqrt{d_1{}^2 - 2d_1 d_2}$$

THREE EQUAL CYLINDERS

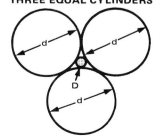

$$D = 0.1547d$$

Rules relative to the circle

Introduction.
This article begins with a common circle and the expressions relating its various parts. From this beginning, the article expands the concept to the relationship between a unit circle and unit square and ends up with the relation between a unit circle and number of regular unit polygons.

The circle is defined as the locus of all the points in a single plane at an equal distance from a given point—the point being the center of the circle. The diameter is the chord of the circle that passes through this center point and coincidently is the longest chord. The circle radius is one-half the diameter and the circumference is the total distance around the perimeter.

Nomenclature:
   d = Diameter
   r = Radius
   A = Area
   C = Circumference

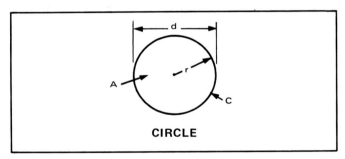

CIRCLE

Beginning with the diameter we now relate all the other parts of a circle to this one part:

$$d = 2r$$
$$d = \frac{C}{\pi}$$
$$d = 0.3183C$$
$$d = \left(\frac{A}{0.7854}\right)^{1/2}$$

The following relationships are relative to the circle radius:

$$r = 0.5d$$
$$r = \left(\frac{A}{\pi}\right)^{1/2}$$
$$r = \frac{C}{2\pi}$$
$$r = \frac{C}{6.28318}$$
$$r = 0.15915C$$

The following relationships are relative to the circle's area:

$$A = \pi r^2$$
$$A = \frac{\pi d^2}{4}$$
$$A = 0.7854d^2$$
$$A = \frac{d}{4}C$$
$$A = 0.07958C^2$$

The following relationships are relative to the circle's circumference:

$C = \pi d$

$C = 2\pi r$

$C = \dfrac{d}{0.3183}$

Circle and square having the same area.

Assume now that we have a circle and square both having the same area. Given this condition we can write the following series of relationships between the two figures:

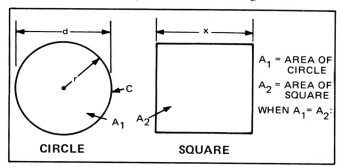

A_1 = AREA OF CIRCLE

A_2 = AREA OF SQUARE

WHEN $A_1 = A_2$:

CIRCLE   **SQUARE**

Circle	Square
$d = 1.1284x$	$x = 1.77246r$
$d = \dfrac{x}{0.8862}$	$x = 0.8862d$
$C = \dfrac{x}{0.2821}$	$x = \dfrac{d}{1.1284}$
$C = 3.545x$	$x = 0.2821C$
$d = \dfrac{3.545x}{\pi}$	$x = \dfrac{C}{3.545}$
$r = 0.56419x$	

Circle and square having the same periphery.

If we have a circle and square both having the same periphery, we find that $d = 1.27324x$. Conversely, if the sum of the four sides of a square is the same length as the circumference of a circle, we find that $x = 0.7854d$.

Also, if we take any regular polygon and let A = area, r = radius of the inscribed circle, n = number of sides and ℓ = length of one side, we obtain the following equation:

$$A = \frac{rn\ell}{2}$$

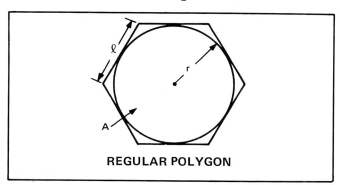

REGULAR POLYGON

Example 1:

Given a circle and square both having the same area, determine the relationship between the radius of the circle (r) and the side of the square (x).

Solution 1:

Going back to the article we find that the following equation applies:

$r = 0.56419x$

REGULAR POLYGONS

SIDES	NAME	AREA (A) WHEN DIAMETER OF INSCRIBED CIRCLE = 1	AREA (A) WHEN SIDE = 1
3	Triangle	1.299	0.433
4	Square	1.000	1.000
5	Pentagon	0.908	1.720
6	Hexagon	0.866	2.598
7	Heptagon	0.843	3.634
8	Octagon	0.828	4.828
9	Nonagon	0.819	6.182
10	Decagon	0.812	7.694
11	Undecagon	0.807	9.366
12	Dodecagon	0.804	11.196

REGULAR POLYGONS

SIDES	NAME	RADIUS (r) OF CIRCUMSCRIBED CIRCLE WHEN SIDE = 1	LENGTH (ℓ) OF SIDE WHEN RADIUS (r) OF CIRCUMSCRIBED CIRCLE = 1
3	Triangle	0.577	1.732
4	Square	0.707	1.414
5	Pentagon	0.851	1.176
6	Hexagon	1.000	1.000
7	Heptagon	1.152	0.868
8	Octagon	1.307	0.765
9	Nonagon	1.462	0.684
10	Decagon	1.618	0.618
11	Undecagon	1.775	0.563
12	Dodecagon	1.932	0.518

REGULAR POLYGONS

SIDES	NAME	LENGTH (ℓ) OF SIDE WHEN PERPENDICULAR TO CENTER = 1	PERPENDICULAR (p) TO CENTER WHEN SIDE = 1
3	Triangle	3.464	0.289
4	Square	2.000	0.500
5	Pentagon	1.453	0.688
6	Hexagon	1.155	0.866
7	Heptagon	0.963	1.038
8	Octagon	0.828	1.207
9	Nonagon	0.728	1.374
10	Decagon	0.650	1.539
11	Undecagon	0.587	1.703
12	Dodecagon	0.536	1.866

International standard prefixes conversion table

Introduction.
The following table of International Standard Prefixes is used to indicate decimal point movement and conversion of units.

Example 1:
Convert 6 microinches to inches.

Solution 1:
Enter the table at micro in the left-hand column and proceed in a horizontal direction until you arrive at the number 6 under the vertical unity column. Now move the decimal point to the left (direction of the arrow) 6 places. Thus the answer is:

6 microinches = 0.000006 inch

Note:
When solving a problem, if you end up on the left side of the shaded diagonal, you observe the shaded arrow and move the required number of places to the left—if you end up on the right side of the shaded diagonal, you observe the shaded arrow and move the required number of places to the right.

QUANTITY	MULTIPLES AND SUBMULTIPLES	EXPLANATION	GIVEN	Symbol	Tera	Giga	Mega	Kilo	Hecto	Deka	UNITY	Deci	Centi	Milli	Micro	Nano	Pico
1 000 000 000 000	10^{12}	One trillion times	Tera	T		3	6	9	10	11	12	13	14	15	18	21	24
1 000 000 000	10^{9}	One billion times	Giga	G	3		3	6	7	8	9	10	11	12	15	18	21
1 000 000	10^{6}	One million times	Mega	M	6	3		3	4	5	6	7	8	9	12	15	18
1 000	10^{3}	One thousand times	Kilo	K	9	6	3		1	2	3	4	5	6	9	12	15
100	10^{2}	One hundred times	Hecto	h	10	7	4	1		1	2	3	4	5	8	11	14
10	10^{1}	Ten times	Deka	da	11	8	5	2	1		1	2	3	4	7	10	13
UNITY			UNITY		12	9	6	3	2	1		1	2	3	6	9	12
0.1	10^{-1}	One tenth of	Deci	d	13	10	7	4	3	2	1		1	2	5	8	11
0.01	10^{-2}	One hundredth of	Centi	c	14	11	8	5	4	3	2	1		1	4	7	10
0.001	10^{-3}	One thousandth of	Milli	m	15	12	9	6	5	4	3	2	1		3	6	9
0.000 001	10^{-6}	One millionth of	Micro	μ	18	15	12	9	8	7	6	5	4	3		3	6
0.000 000 001	10^{-9}	One billionth of	Nano	n	21	18	15	12	11	10	9	8	7	6	3		3
0.000 000 000 001	10^{-12}	One trillionth of	Pico	p	24	21	18	15	14	13	12	11	10	9	6	3	

Coordinates of parabolic curve

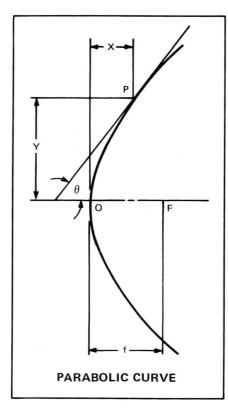

PARABOLIC CURVE

Introduction.
Frequently in engineering calculations it is necessary to determine the X and Y coordinates of a point P located on a parabolic curve that has a focal length f. The nomogram can also be used to find the angle θ of the tangent line through point P.

Example 1:
Given a parabola that has an f = 1.0 inch and Y coordinate of 1.5 inches, determine the X coordinate.

Solution 1:
Construct a line from 1.0 on the focal length scale to 1.5 on the Y coordinate scale and continue this line until it intersects the X coordinate scale line. At this intersection point read the answer of X = 0.56 inches.

Now, construct a line parallel to the above line passing through Point A and intersecting the angle of tangent line scale and read the answer of $\theta = 53$ deg.

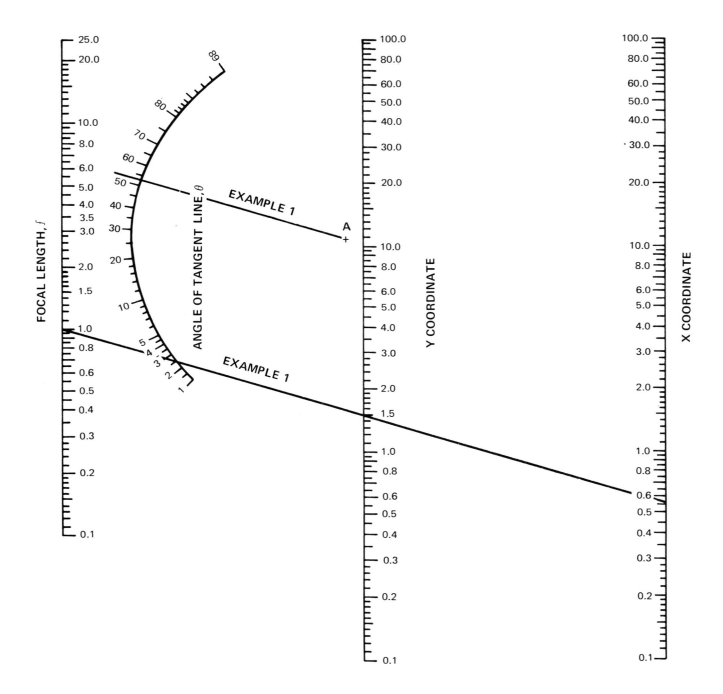

Decimal-fraction nomogram

Introduction.
The following nomogram facilitates the addition and subtraction of numbers containing fractions with those containing decimals. Addition and subtraction of decimals or fractions is also simplified.

Example 1:
Given a decimal of 6.43 and a fraction of 1-49/64, determine their sum.

Solution 1:
Using the addition equation:

$$A + B = C$$
$$6.43 + 1\text{-}49/64 = C$$
$$(6 + 1) + 0.43 + 49/64 = C$$

Using the nomogram, construct a line from 0.43 on the decimals, A scale to 49/64 on the fractions, B scale. The answer can be read from the decimals, C scale or fractions, C scale:

$$7 + 1.187 = 8.187$$
$$\text{or}$$
$$7 + 1\text{-}3/16 = 8\text{-}3/16$$

Example 2:
Given a fraction of 11-7/8 and a decimal of 3.53, determine their difference.

Solution 2:
Using the subtraction equation:

$$C - B = A$$
$$11\text{-}7/8 - 3.53 = A$$
$$(11 - 3) + 7/8 - 0.53 = A$$

Using the nomogram, construct a line from 7/8 on the fractions, C scale to 0.53 on the decimals, B scale. Continuing this line, the answer can be read from the fractions, A scale or decimals, A scale:

$$8 + 11/32 = 8\text{-}11/32$$
$$\text{or}$$
$$8 + 0.343 = 8.343$$

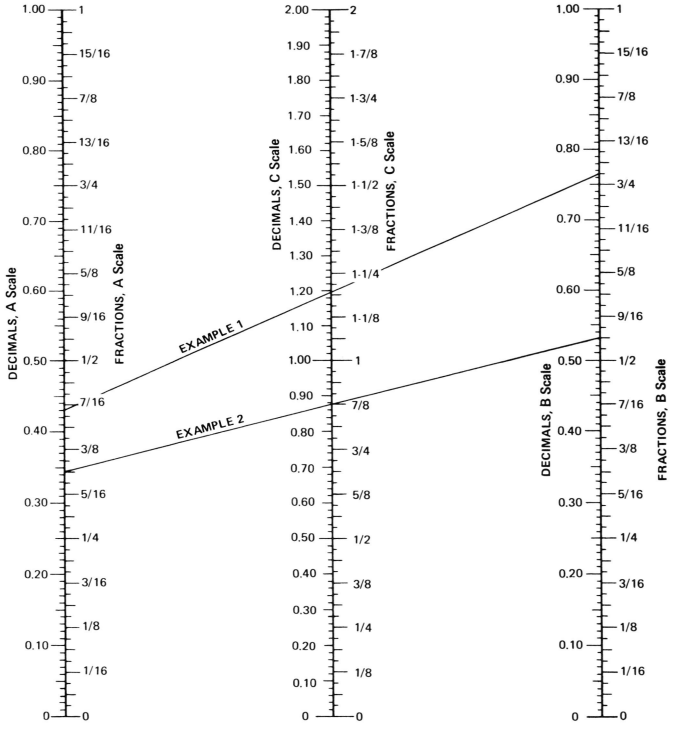

Solution of geometric problems

Introduction.

The following article contains the most common geometric problems and solutions that frequently confront the engineer. The drawings are self-explanatory and the known or measured quantities are stated and then the desired results that are to be found are indicated. All dashed lines shown are construction lines to aid in showing how the solution was obtained.

Example 1:

Given a triangle that has $\beta = 40°$, $\gamma = 35°$ and b = 8.0 inches, determine side c.

Solution 1:

Using the solution of Fig 1:

$$\frac{b}{\text{Sin }\beta} = \frac{c}{\text{Sin }\gamma}$$

$$\frac{8.0}{\text{Sin }40°} = \frac{c}{\text{Sin }35°}$$

$$\frac{8.0}{0.643} = \frac{c}{0.574}$$

$$0.643c = (8.0)(0.574)$$

$$c = \frac{(8.0)(0.574)}{0.643}$$

$$c = 7.1 \text{ inches}$$

Given: Two sides and angle opposite one of them
Solution: Law of sines

$$\frac{a}{\sin\alpha} = \frac{b}{\sin\beta} = \frac{c}{\sin\gamma}$$

Given: Two sides and included angle or all three sides
Solution: Law of cosines

$$a^2 + b^2 - c^2 - 2ab\cos\gamma = 0$$
$$a^2 - b^2 + c^2 - 2ac\cos\beta = 0$$
$$-a^2 + b^2 + c^2 - 2bc\cos\alpha = 0$$

Angles α, β and γ are respectively opposite sides a, b and c.

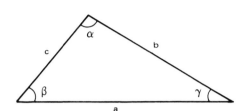

Fig. 1—Solution of Triangles

Given: e and K, determine α
Solution:

$$\sin\alpha = \frac{e}{k}$$

Fig. 2—Sine Bar

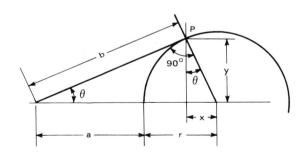

Given: a and r, determine x and y
Solution:

$$b^2 = (a + r)^2 - r^2$$
$$\sin\theta = \frac{r}{a+r} = \frac{y}{b} = \frac{x}{r}$$
$$x = \frac{r^2}{a+r}$$
$$y = \frac{br}{a+r}$$

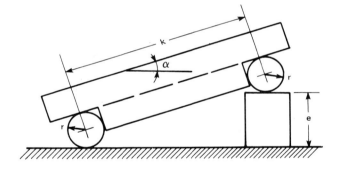

Fig. 3—Determination of Point of Tangency with Arc

Given: b, c and r, determine x_1 and y_1
Solution:

$$a = \sqrt{b^2 + c^2} - r$$

$$\sin \theta = \frac{r}{a + r}$$

$$\tan \alpha = \frac{b}{c}$$

$\gamma = 90° - \theta - \alpha$
$e = r \sin\gamma$, then $x_1 = b + e$
$d = r \cos\gamma$, then $y_1 = c - d$

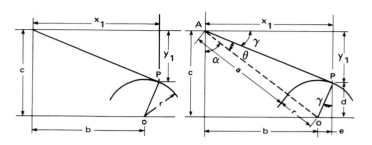

Fig. 4—Determination of Point of Tangency with Circle

Given: a, b, r_1 and r_2, determine x_1, y_1, x_2 and y_2
Solution:

$$c = \sqrt{a^2 + b^2}$$

$$\tan \theta = \frac{b}{a}$$

$$\sin \phi = \frac{r_2 - r_1}{c}$$

$x_1 = r_1 \sin (\theta + \phi)$
$y_1 = r_1 \cos (\theta + \phi)$
$x_2 = r_2 \sin (\theta + \phi)$
$y_2 = b + r_2 \cos (\theta + \phi)$

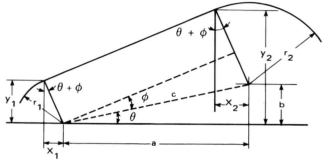

Fig. 5—Layout and Equations for Finding Points of Tangency with Circles

---

Given: a, b, d and e, determine x and y
Solution:

$$c^2 = d^2 + e^2$$

$$\cos \alpha = \frac{b^2 + c^2 - a^2}{2bc}$$

$$\tan \theta = \frac{e}{d}$$

$x = b \cos (\alpha + \theta)$
$y = b \sin (\alpha + \theta)$

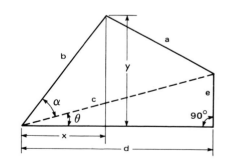

Fig. 6—Equations for Finding Coordinates of Point

Given: a, b, r_1 and d_2, determine θ and x
Solution:

$$\sin \theta = \frac{2r_1 + d_2}{a_1 - b_1}$$

$$x = \frac{1}{2}(a_1 + b_1) + r_1 \cot \theta$$

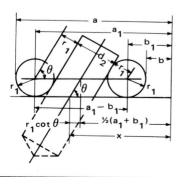

Fig. 7—Equations for Finding Location and Angle of Inclined Hole

Given: L, r and h, determine R
Solution:

$$(R + r)^2 = (R + h - r)^2 + \left(\frac{1}{2} L\right)^2$$

$$R = \frac{L^2}{8(2r - h)} - \frac{h}{2}$$

Given: L, r and h_1, determine R
Solution:

$$(R - r)^2 = (R - h_1 + r)^2 + \left(\frac{1}{2} L\right)^2$$

$$R = \frac{L^2}{8(h_1 - 2r)} + \frac{h_1}{2}$$

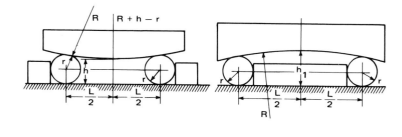

Fig. 8—Equations for Finding Length of Long Radius

Given: a, b, c, d, r_1 and r_2, determine x, y, α and β
Solution:

$$\tan \phi = \frac{b_1 - a_1}{d_1 - c_1}$$

$$\overline{0_1 0_2} = \frac{d_1 - c_1}{\cos \phi}$$

$$\sin \theta = \frac{r_2 - r_1}{\overline{0_1 0_2}}$$

$$\overline{00_1} = \frac{r_1}{\sin \theta}$$

$$x = a_1 - \overline{00_1} \sin \phi$$

$$y = c_1 - \overline{00_1} \cos \phi$$

$$\alpha = 90° + \phi - \theta$$
$$\beta = 90° - \phi - \theta$$

Special case for $\phi = 0$, then:

$$\alpha = \beta$$

$$\overline{0_1 0_2} = d_1 - c_1$$

$$\sin \theta = \frac{r_2 - r_1}{d_1 - c_1}$$

$$y = c_1 - \frac{r_1}{\sin \theta}$$

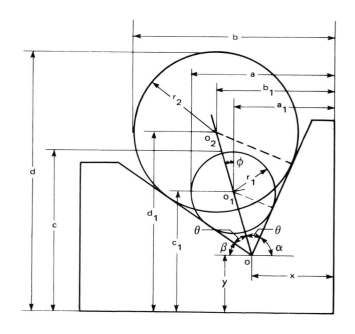

Fig. 9—Equations for Measuring V-notch

195

Given: a, b, r_1 and r_2, determine x and ϕ
Solution:

$$\tan \theta = \frac{r_2 - r_1}{b_1 - a_1}$$

$$\phi = 90° - 2\theta$$
$$x = a_1 - r_1 \cot \theta$$

Given: a, b, r and h, determine x and ϕ
Solution:

$$\tan 2\theta = \frac{h}{b - a}$$

$$\phi = 90 - 2\theta$$
$$x = a - r - r \cot \theta$$

Given: c, d, r_1 and r_2, determine x and ϕ
Solution:

$$\sin \phi = \frac{r_2 - r_1}{d_1 - c_1}$$

$$x = \frac{r_1}{\cos \phi} - c_1 \tan \phi$$

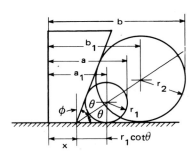
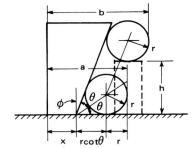
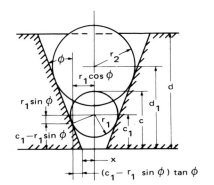

Fig. 10—Equations for Measuring External Taper

Given: b, c, d, r_1 and r_2, determine C, D, ϕ and R_1
Solution:

$$C^2 = 2c(r_1 + r_2) - c^2$$
$$D^2 = 2d(r_1 + r_2) - d^2$$

$$\tan \theta = \frac{D - C}{b}$$

$$2\theta = 90° + \phi$$
$$R_1 = C + r_1 \cot \theta$$

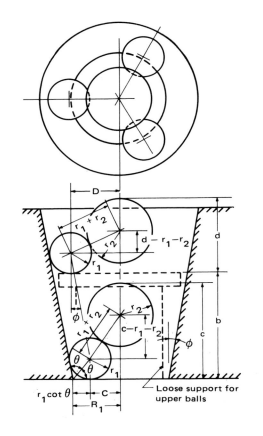

Fig. 11—Layout and Equations for Finding Internal Tapers

Loose support for upper balls

196

Perimeter of ellipse

Introduction.

Frequently in engineering calculations it is necessary to calculate the perimeter of an ellipse. The following table expedites that process.

Nomenclature:

P = Perimeter of ellipse

A = Major axis of ellipse

B = Minor axis of ellipse

$f\left(\dfrac{B}{A}\right)$ = Function of $\dfrac{B}{A}$

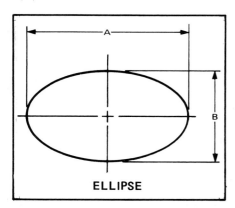

ELLIPSE

The perimeter of an ellipse may be calculated from the following equation:

$$P = Af\left(\frac{B}{A}\right) \qquad (1)$$

Example 1:

The major axis of an ellipse is 5.0 inches and the minor axis is 2.0 inches, determine its perimeter.

Solution 1:

Solving for $\dfrac{B}{A}$:

$$\frac{B}{A} = \frac{2.0}{5.0} = 0.40$$

From the table we find that $f\left(\dfrac{B}{A}\right) = 2.3013$

Substituting back into Eq. 1:

$$P = Af\left(\frac{B}{A}\right)$$

$$P = 5.0(2.3013)$$

$$P = 11.5 \text{ inches}$$

$\dfrac{B}{A}$	$f\left(\dfrac{B}{A}\right)$	$\dfrac{B}{A}$	$f\left(\dfrac{B}{A}\right)$
0.00	2.0000	0.51	2.4348
0.01	2.0008	0.52	2.4475
0.02	2.0021	0.53	2.4604
0.03	2.0041	0.54	2.4733
0.04	2.0066	0.55	2.4863
0.05	2.0098	0.56	2.4994
0.06	2.0134	0.57	2.5126
0.07	2.0174	0.58	2.5259
0.08	2.0219	0.59	2.5393
0.09	2.0268	0.60	2.5527
0.10	2.0320	0.61	2.5662
0.11	2.0376	0.62	2.5798
0.12	2.0435	0.63	2.5935
0.13	2.0497	0.64	2.6072
0.14	2.0562	0.65	2.6211
0.15	2.0631	0.66	2.6349
0.16	2.0701	0.67	2.6489
0.17	2.0775	0.68	2.6629
0.18	2.0851	0.69	2.6770
0.19	2.0929	0.70	2.6912
0.20	2.1010	0.71	2.7054
0.21	2.1093	0.72	2.7197
0.22	2.1178	0.73	2.7341
0.23	2.1266	0.74	2.7485
0.24	2.1355	0.75	2.7629
0.25	2.1446	0.76	2.7775
0.26	2.1539	0.77	2.7921
0.27	2.1634	0.78	2.8067
0.28	2.1731	0.79	2.8214
0.29	2.1829	0.80	2.8362
0.30	2.1930	0.81	2.8510
0.31	2.2031	0.82	2.8658
0.32	2.2135	0.83	2.8808
0.33	2.2240	0.84	2.8957
0.34	2.2346	0.85	2.9108
0.35	2.2454	0.86	2.9258
0.36	2.2563	0.87	2.9409
0.37	2.2673	0.88	2.9561
0.38	2.2785	0.89	2.9713
0.39	2.2899	0.90	2.9866
0.40	2.3013	0.91	3.0019
0.41	2.3129	0.92	3.0172
0.42	2.3246	0.93	3.0326
0.43	2.3364	0.94	3.0481
0.44	2.3483	0.95	3.0636
0.45	2.3603	0.96	3.0791
0.46	2.3725	0.97	3.0946
0.47	2.3847	0.98	3.1103
0.48	2.3971	0.99	3.1259
0.49	2.4096	1.00	3.1416
0.50	2.4221		

Squares, square root graph

Introduction.
Frequently, the square and/or square root of a number is required. The following graph provides a quick solution to these types of problems.

Example 1:
Given the number 5, determine the square of it.

Solution 1:
Locate the number 5 on the left side of the scale and directly read the square of 5 on the right side of the scale as 25.

Example 2:
Given the number 36, determine the square root of it.

Solution 2:
Locate the number 36 on the right side of the scale and directly read the square root of 36 on the left side of the scale as 6.

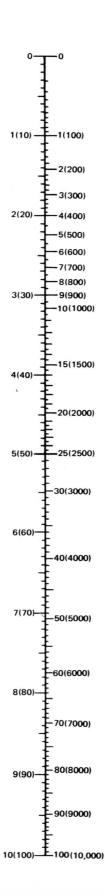

Cubes, cube root graph

Introduction.
Frequently, the cube and/or cube root of a number is required. The following graph provides a quick solution to these types of problems.

Example 1:
Given the number 2, determine the cube of it.

Solution 1:
Locate the number 2 on the right side of the scale and directly read the cube of 2 on the left side of the scale as 8.

Example 2:
Given the number 64, determine the cube root of it.

Solution 2:
Locate the number 64 on the left side of the scale and directly read the cube root of 64 on the right side of the scale as 4.

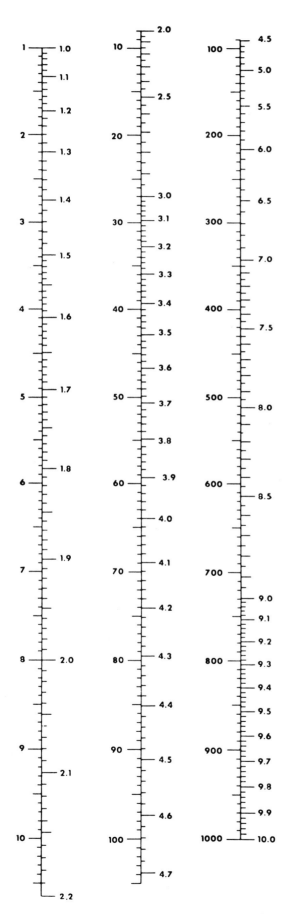

Coordinates for equally spaced holes

Introduction.

It is sometimes necessary to determine the x and y coordinates of a circle divided into an equal number of parts. The following table can be used directly during initial layout or it can serve as a crosscheck against answers obtained by normal trigonometric methods.

Example 1:

Given a circle that has a radius of 5.0 inches and contains 4 holes spaced at 90°, determine the distance between their centers.

Solution 1:

Going to the table we locate our particular case. Substituting into the equation:

A = 1.4142R
A = 1.4142(5.0)
A = 7.07 inches

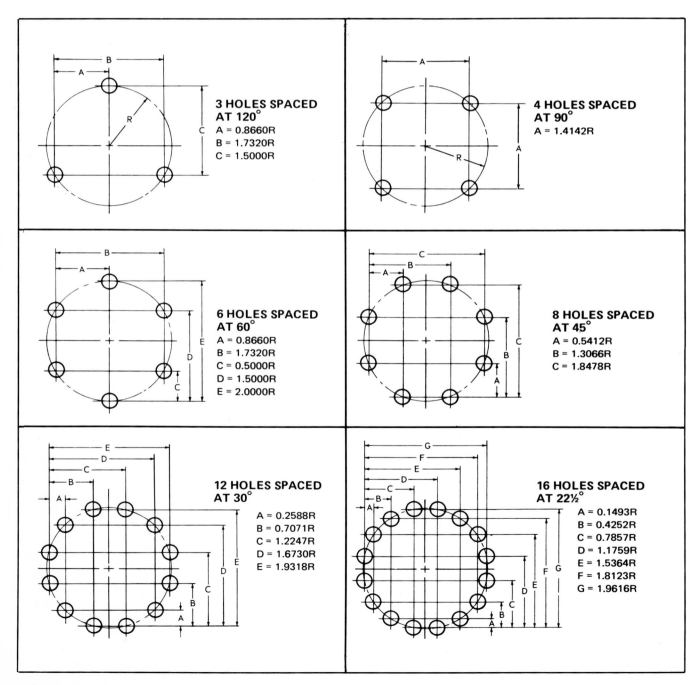

3 HOLES SPACED AT 120°
A = 0.8660R
B = 1.7320R
C = 1.5000R

4 HOLES SPACED AT 90°
A = 1.4142R

6 HOLES SPACED AT 60°
A = 0.8660R
B = 1.7320R
C = 0.5000R
D = 1.5000R
E = 2.0000R

8 HOLES SPACED AT 45°
A = 0.5412R
B = 1.3066R
C = 1.8478R

12 HOLES SPACED AT 30°
A = 0.2588R
B = 0.7071R
C = 1.2247R
D = 1.6730R
E = 1.9318R

16 HOLES SPACED AT 22½°
A = 0.1493R
B = 0.4252R
C = 0.7857R
D = 1.1759R
E = 1.5364R
F = 1.8123R
G = 1.9616R

References

1. "Angular Relationships," DESIGN NEWS, Boston, MA, Aug. 5, 1974.
2. Apalategi, J.J. and Adams, L.J., "Aircraft Analytic Geometry," Chapter 2: Plane Analytic Geometry, pgs. 20-21; Chapter 11: Conics, Analytical Theory, The Parabola, pgs. 235-237, McGraw-Hill Book Co., New York, NY, 1944.
3. Bartee, E.M., "Engineering Experimental Design Fundamentals," pg. 58, Prentice-Hall, Inc., Englewood Cliffs, NJ, 1968.
4. Bowker, A.H. and Lieberman, C.J., "Engineering Statistics," pgs. 144-147, Prentice-Hall, Inc., Englewood Cliffs, NJ, 1959.
5. Burlington, Richard S., "Handbook of Mathematical Tables and Formulas," Handbook Publishers Inc., Sandusky, OH, 3rd Edition, 1948, Chapter 1, pgs. 1-10.
6. Cone, Irwin, "Squares and Square Root Chart," DESIGN NEWS, Denver, CO, Oct. 1, 1956.
7. Costa, David, "Geometrical Curves for Reference," DESIGN NEWS, Denver, CO, Nov. 13, 1963.
8. "Cube Roots of Numbers," DESIGN NEWS, Denver, CO, Feb. 1, 1957.
9. "Decimal-Fraction Nomogram," DESIGN NEWS, Denver, CO, Feb. 15, 1956.
10. "Degree-Radian Comparisons," DESIGN NEWS, Denver, CO, Mar. 20, 1963.
11. De Groot, A., "Engineering Mechanics," Part 1: Plane Rectangular Coordinates, pgs. 27-28, International Textbook Co., Scranton, PA, 1951.
12. Dixon, W.J. and Massey, F.J., "Introduction to Statistical Analysis," 2nd Edition, pg. 124, McGraw-Hill Book Co., New York, 1957.
13. Dolciani, Mary P.; Berman, Simon L.; Wooton, William; "Modern Algebra and Trigonometry," Chapter 1: Sets of Numbers, Axioms, Theorems, pgs. 8-9, 19-22; Chapter 2: Corollary, pgs. 69-70, Houghton Miffin Co., 1965.
14. Douglass, Raymond D. and Zeldin, Samuel D., "Calculus and Its Applications," pgs. 148, 221 and 335, Prentice-Hall, Inc., New York, 1947.
15. Feingold, Samuel, "A Table Of Fillet Areas," DESIGN NEWS, Denver, CO, Jan. 20, 1969.
16. Feingold, Samuel, "Coordinates for Equally Spaced Holes," DESIGN NEWS, Denver, CO, May 12, 1969.
17. "Finding Areas of a Few Common Shapes," DESIGN NEWS, Denver, CO, Mar. 1, 1967.
18. Gheorghiu, Paul, "Nomogram for Solving Equations of the Form $c = \sqrt{a^2 + b^2}$," DESIGN NEWS, Denver, CO, Apr. 25, 1960.
19. Gheorghiu, Paul, "Nomogram for Solving Equations of the Form $Z = X^y$," DESIGN NEWS, Denver, CO, July 4, 1960.
20. Grabill, James R., "Graphic Aid for Cubic Equations," DESIGN NEWS, Denver, CO, Mar. 6, 1963.
21. Guttman, I. and Wilks, S.S., "Introductory Engineering Statistics," pgs. 34-37, John Wiley & Sons, Inc., New York, 1965.
22. Hahn, G.J. and Shapiro, S.S., "Statistical Models in Engineering," pgs. 110-112, John Wiley & Sons, Inc., New York, 1967.
23. Hart, William L.; Wilson, W.A.; Tracey, J.I.; "First Year College Mathematics," Chapter 1: Rectangular Cartesian Coordinates, pgs. 2-9; Chapter 3: The Straight Line, pgs. 41-43; Chapter 6: The Ellipse, pgs. 92-94, D.C. Heath & Co., 1943.
24. Hodska, Nicholas, "Basic Laws of Algebra," DESIGN NEWS, Boston, MA. July 8, 1975.
25. Hodska, Nicholas, "Basic Laws of Geometry," DESIGN NEWS, Boston, MA. July 28, 1975.
26. Hodska, Nicholas, "Course on Geometric and Mensuration Relationships," DESIGN NEWS, Boston, MA, 1974.
27. Hodska, Nicholas, "Tables of Differential and Integral Calculus," DESIGN NEWS, Boston, MA, Aug. 19, 1975.
28. Hooke, Robert and Shaffer, Douglas, Math and Aftermath, "Appendix: A Brief Review of Calculus," pgs. 204-205, Walker & Co., New York, 1965.
29. Hutton, Charles and Rutherford, William, "A Course of Mathematics," William Tegg and Co., Cheapside, London, England; Printed by James Nichols, Hoxton Square, London, England; Chapter on "Geometry of Planes," pgs. 417-419; Chapter on "Solid Geometry" and "Spherical Geometry," pgs. 431-442.
30. Hutton, Charles and Rutherford, William, "A Course in Mathematics," Chapter on Geometry: Definitions, pgs. 357-359; Theorems, pgs. 362-372; William Tegg & Co., London, 1850.
31. "International Standard Prefixes," DESIGN NEWS, Denver, CO, Sept. 1, 1961.
32. Johnson, N.L. and Leone, F.C., "Statistics and Experimental Design in Engineering and Physical Sciences," Vols. I and II, pgs. 67-70 and pgs. 84-87,

References

John Wiley & Sons, Inc., New York, 1964.

33. Kaye, S. Warren, "Simplified Solution of Hyperbolic Functions and Exponentials," DESIGN NEWS, Denver, CO, Nov. 15, 1957.

34. Kimball, R.E., "Practical Mathematics," National Educational Alliance Inc., New York, NY, Volume III, 1945, "Strange Ways With Figures," pg. 913.

35. Kindle, J.H., "Theory and Problems of Plane and Solid Analytic Geometry," Chapter 10: Tangents and Normals, pgs. 84-88, Schaum Publishing Co., New York, NY, 1950.

36. Kravitz, Sidney, "Cones with Rounded Tips," DESIGN NEWS, Denver, CO, Feb. 16, 1966.

37. Kravitz, Sidney, "Constants Containing π," DESIGN NEWS, Denver, CO, July 24, 1963.

38. Kravitz, Sidney, "Ellipse Perimeters," DESIGN NEWS, Denver, CO, Mar. 3, 1965.

39. Kravitz, Sidney, "Fitting Circular Cross-Sections in Other Configurations," DESIGN NEWS, Denver, CO, Dec. 9, 1964.

40. Kravitz, Sidney, "Tapers, Slopes and Their Angles," DESIGN NEWS, Denver, CO, July 10, 1963.

41. Lowe, Howell, "Graphical Method for Fundamental Trigonometric Relations," DESIGN NEWS, Denver, CO, Dec. 15, 1955.

42. Machinery Handbook, "Positive and Negative Numbers" and "Geometrical Progressions," The Industrial Press, New York, NY, 15th Edition, 1956, pgs. 105, 270-275.

43. Marks, Lionel S., "Mechanical Engineers' Handbook," McGraw-Hill Book Co., New York, NY, 5th Edition, 1951, Section 2: Mathematics by Edward Huntington and Philip Franklin, "Geometry and Mensuration," pgs. 97-109; "Geometrical Theorems," pgs. 97-99; Derivatives and Differentials," pgs. 150-151; "Indefinite Integrals," pgs. 156-159.

44. "Nomogram for Law of Cosines," DESIGN NEWS, Boston, MA, Dec. 4, 1972.

45. "Nomogram for Law of Sines," DESIGN NEWS, Boston, MA, Oct. 9, 1972.

46. Oberg, Erik and Jones, F.D., "Machinery's Handbook," The Industrial Press, New York, NY, 15th Edition, 1956, Chapters on "Areas and Volumes" and "Mensuration," pgs. 148-162.

47. Patel, S.M., "Volume of Rings," DESIGN NEWS, Boston, MA, Sept. 3, 1973.

48. Peck, William G., "A Treatise on Analytic Geometry," The Circle, pgs. 42-45; the Hyperbola, pgs. 122-126; A.S. Barnes & Co., New York, NY, 1873.

49. Peters, Robert L., "Quadratic Equation Nomograms," DESIGN NEWS, Denver, CO, Mar. 17, 1965.

50. "Polar to Rectangular Coordinates Nomogram," DESIGN NEWS, Boston, MA, Feb. 19, 1973.

51. Post, William, "Decimal Degree Equivalents," DESIGN NEWS, Boston, MA, Apr. 17, 1972.

52. Saelman, B., "Distance Between Two Skew Lines," DESIGN NEWS, Boston, MA, Aug. 21, 1972.

53. Saelman, B., "Short Method of Trigonometric Solutions," DESIGN NEWS, Denver, CO, Sept. 15, 1955.

54. Saelman, B., "Short Method for Hyperbolic Solutions," DESIGN NEWS, Denver, CO, July 1, 1956.

55. Seman, Michael C., "Simplified Solution of Triangles," DESIGN NEWS, Denver, CO, May 24, 1971.

56. Smith, Bart A., "Notes on the Development of a Symmetrical Cone," DESIGN NEWS, Denver, CO, Feb. 16, 1968.

57. Smith, Bart A., "Rules Relative to The Circle," DESIGN NEWS, Denver, CO, Aug. 5, 1968.

58. Spotts, M.F., "Equation for Finding the Volume of Solids," DESIGN NEWS, Boston, MA, Oct. 8, 1973.

59. Spotts, M.F., "Numerical Integration for Area Under a Curve," DESIGN NEWS, Boston, MA, Aug. 18, 1975.

60. Spotts, M.F., "Numerical Solution of Complex Equations," DESIGN NEWS, Boston, MA, 1975.

61. Spotts, M.F., "Solution of Geometric Problems," DESIGN NEWS, Boston, MA, Feb. 3, 1975.

62. Spregel, M.R., "Theory and Problems of Statistics," pg. 46, Schaum Publishing Co., New York, 1961.

63. Stephenson, James, "Fillet Area (any angle fillet)," DESIGN NEWS, Boston, MA, May 5, 1975.

64. "Trig Function Reference Circle," DESIGN NEWS, Denver, CO, Apr. 1, 1964.

65. Wadler, Richard, "Coordinates of a Point on a Parabolic Curve," DESIGN NEWS, Denver, CO, Mar. 27, 1961.

66. Zanker, Adam, "Definite Integral Nomogram," DESIGN NEWS, Boston, MA, Nov. 5, 1973.

67. Zanker, Adam, "Solutions for the Unsolvable Equations," DESIGN NEWS, Boston, MA, Apr. 7, 1975.